"Oné Pagán is a brilliant biologist and a charming storyteller. Just like his other works, *Drunk Flies and Stoned Dolphins* will make you look at the world with awestruck eyes."

—Matthew D. LaPlante, author of *Superlative*

"Drugs—illicit or otherwise—have an impact on us. From coffee to cocaine, drugs affect our thinking, our moods, and our actions. But is this a uniquely human trait? In *Drunk Flies and Stoned Dolphins*, Oné Pagán explores the impact of our pharmaceutical cocktails on the natural world and in so doing teaches us a little more about ourselves. As remarkable as humans are, we're not distinct from animals. We are an intricate part of the animal kingdom. Our experience with drugs is not that dissimilar to those of other animals and sometimes the results are quite hilarious. Santa Claus and magic mushrooms? Squirrels on amphetamines? Dolphins on LSD and the Search for Extraterrestrials? Birds grabbing cigarette butts for their nests? *Drunk Flies and Stoned Dolphins* has it all! If you're looking for a great science read, I highly recommend *Drunk Flies and Stoned Dolphins*."

—Peter Cawdron, author of *3zekiel*

"*Drunk Flies and Stoned Dolphins* was an absolute joy and pleasure to read. This book is the whole package, it has incredible science, unexpected concepts, fun facts, great personal stories, and awesome dad jokes that had me interested and laughing throughout the book (please don't miss the footnotes). Dr. Pagán is an extraordinary science communicator and I love how meticulous he is on his writing, delivering a compelling story; it actually feels like having a conversation with him. If you are interested in how animals and other living creatures behave around drugs, this book is for you."

—Sofía Villalpando, biologist, science communicator, entrepreneur, and blogger at The Biologist Apprentice

"With a down-to-earth conversational style and witty sense of humor, Dr. Oné Pagán brings the quirks of the animal kingdom to life with his writing. *Drunk Flies and Stoned Dolphins* showcases the fascinating world of psychoactive substances as experienced by organisms ranging from tiny microbes to humans."

—Marie McNeely, PhD, host of *The People Behind the Science* podcast

"With endless stories of animals—from fruit flies to elephants—reacting in such human-like ways to intoxicating substances, Pagán makes good on his promise to show that animals are not so different from us when it comes to the desire to alter our minds. Pagán presents what could be dense material (including plenty of chemistry) in easy to digest explainers and anecdotes, yet keeps the book tethered to the hard-nosed scientific literature in a way that will satisfy the most curious—and skeptical—of minds. Pagán's writing style is breezy and conversational, which makes for a pleasurable read. And please don't skip over the footnotes: Pagán's playful personality shines through in these hidden gems!"

—Justin Gregg, PhD, senior research associate with the Dolphin Communication Project, adjunct professor in the Department of Biology at St. Francis Xavier University, and author of the book *Are Dolphins Really Smart?*

"Dr. Oné Pagán is one of those few brilliant modern science writers and educators who can capture the charm and curiosity of yesteryear, yet whose writing is imbued with the poise and sensitivity requisite of the 21st century, all crafted through the lens of purely masterful storytelling. While illuminating the murkier areas of biology, *Drunk Flies and Stoned Dolphins* also provides ample evidence that the future of science writing is in *very* good hands."

—Micah Hanks, writer, journalist, podcaster, and creator of *The Debrief*

# DRUNK FLIES
## AND STONED
## DOLPHINS

## ALSO BY ONÉ R. PAGÁN

*The First Brain*
*Strange Survivors*

# DRUNK FLIES AND STONED DOLPHINS

## A TRIP THROUGH THE WORLD OF ANIMAL INTOXICATION

## ONÉ R. PAGÁN

BenBella Books, Inc.
Dallas, TX

BenBella

BenBella Books, Inc.
10440 N. Central Expressway
Suite 800
Dallas, TX 75231
benbellabooks.com
Send feedback to feedback@benbellabooks.com

*BenBella* is a federally registered trademark.

Printed in the United States of America
10 9 8 7 6 5 4 3 2 1

Library of Congress Control Number: 2021941518
ISBN 9781950665372
eISBN 9781950665600

Editing by Laurel Leigh and Alexa Stevenson
Copyediting by Scott Calamar and Michael Fedison
Proofreading by Jenny Bridges and Sarah Vostok
Indexing by Amy Murphy
Text design and composition by Aaron Edmiston
Cover design by Sarah Avinger
Cover photo of Indo-Pacific bottlenose dolphin with pufferfish © Alamy / Blue Planet
    Archive EP-TMI
Back cover photos © Shutterstock / Roblan, Tomasz Klejdysz, and nechaevkon (fruit
    flies) and Unsplash / Quaritsch Photography (cherry)
Printed by Lake Book Manufacturing

---

**Special discounts for bulk sales are available.**
**Please contact bulkorders@benbellabooks.com.**

*To all the scientists and physicians who, following in the footsteps of Edward Jenner, dedicate their research lives to the invention and development of vaccines.*

# CONTENTS

CHAPTER 5

## SPINELESS MINDS ON DRUGS    151

*In which we talk about drunk insects (including one of our title characters), spiders high on drugs, anxious and addicted flatworms, and some chemically-mellowed octopi, among other fascinating examples that showcase one of this book's chief philosophical tenets: that we are not so different from even the smallest animals.*

CHAPTER 6

## BIGGER MINDS ON DRUGS    197

*In which we talk about some of our closer relatives in the animal world, and see what happens when they are on drugs, whether because we administered them (three words: elephants on acid) or because they sought the experiences themselves (goats getting jittery on coffee beans). Oh, and don't forget the (apparently) stoned dolphins . . .*

# AUTHOR'S NOTE

I would like to begin by thanking you, dear reader, for choosing to read this book. As an avid reader myself, I know only too well that there are "so many books, so little time." And yet here we are!

Most of us have direct knowledge or experience of the effects of psychoactive substances—drugs—on people. What's more, there are a plethora of stories on the subject. For that very reason, I wanted to explore the topic from a different angle: the premise that, when it comes to psychoactive substances, we humans are merely the supporting cast. This is what *Drunk Flies and Stoned Dolphins* is about.

The stories I will tell you in this book are by turns surprising, revealing, odd, and thought-provoking, and yet many are also, frankly, quite funny. Therefore, I want you to know two important things: First, I in no way intend to ignore, dismiss, minimize, or make a mockery of drug abuse and addiction, or the considerable pain and suffering these phenomena cause in our society.[1] Second, I am in no way, not even a little, advocating for the consumption of drugs, whether they be legal or otherwise. And please note that I am a scientist, not a physician!

I do indeed hope to entertain you with interesting stories of animals and drugs, but I also hope to make you think, to contribute to our collective knowledge on this topic, and perhaps even help people along the way. Sometimes, by looking at animals, we find it a bit easier to retain the distance needed for logical consideration of a subject, especially when the subject is one (like drug use) that is controversial or emotionally fraught. Information really *is* power. The more we know about the causes of drug abuse and related behaviors, the better position we are in to understand them and to help alleviate their negative effects.

I've written this book in a particular style, as if you and I were having a conversation. In other words, I will be talking *with* you, not *at* you or *past* you; certainly I will never (God forbid!) *lecture* you. But despite this, and the fact that I include anecdotal stories, I do not engage in tall tales; I've documented these stories to the very best of my abilities. To do so, I used published, verified sources such as personal communications from qualified scientists, scientific books, scientific reviews, and original scientific papers.* An invaluable resource in many things psychopharmacological (in human and nonhuman organisms alike) was the late Dr. Ronald K. Siegel's book *Intoxication: The Universal Drive for Mind-Altering Substances*.[2] (Dr. Siegel was an interesting character to say the least, and we will hear more about him and his work later on.) Another important and very entertaining resource was the book *Animals and Psychedelics: The Natural World and the Instinct to Alter Consciousness*, written by Dr. Giorgio Samorini, a prolific researcher and writer on the ethnobotany and anthropology of psychoactive drugs.[3] These are only two resources out of many, and I have included a comprehensive bibliography at the end of the book should you wish to know more about any particular topic. You see, as much as I would like to tell you *everything*, no book can do that. Any honest scientist will admit as much; this is one of the most endearing characteristics of nature as seen through the lens of science. If a scientist, no matter how distinguished, tells you that a particular book is absolutely complete or the definitive account of a particular topic, such a claim usually reveals a profound misunderstanding of what science is—and of what nature can have in store for us.

---

* *A note on websites:* The internet is a dynamic "place." Links are activated, deactivated, and changed every single day. Any links that I refer to in this book were active and displayed the intended information at the time of this writing. That is all anyone can say!

Now, I invite you to join me on a journey into the minds of some of the myriad species who share this marvelous earth with us, a journey that I firmly believe is no less awe-inspiring than a journey to the stars.

–Oné R. Pagán
Somewhere in Pennsylvania
June 2021

# INTRODUCTION

One memorable gift of the early 1980s is the movie *Caveman*, released in 1981 with an iconic cast that included Ringo Starr, Shelley Long, Dennis Quaid, and Barbara Bach. Although not favorably reviewed by film critics, it is one of my favorite movies. The very best part (in my opinion) is when some of the story's heroes are stalked by a hungry-looking dinosaur that very much resembles a *Tyrannosaurus rex*.* This is the most amusing dinosaur I have ever seen on film, bar none—truly, his (or her†) facial expressions are priceless. (In fact, at least one critic at the time of the movie's release considered the dinosaur the *real* star of the show.[1]) Anyway, our heroes try to fend off the terrible lizard with a large, weird shrub that looks like a cross between a poppy and a marijuana plant.‡ Strangely enough, especially in light of the fact that *T. rex* was without a doubt a meat-eater, the dinosaur eats the shrub. And after a few short moments, the big guy's mood noticeably . . . mellows. In fact, he gives the distinct impression of being intoxicated, and of course, hilarity ensues. The mighty king of the Cretaceous stumbles around for a bit, and eventually falls off a cliff. Fortunately, he is apparently unharmed by this, as in a scene of denouement we see the dinosaur (who by now feels like an old friend) sitting on some rocks, displaying a contemplative—almost

---

* Here I must emphatically state that dinosaurs and people never coexisted in real life!

† There is nothing in the movie indicating the sex or gender of the dinosaur; for simplicity's sake, I hope you won't mind if I refer to this wonderful character as he/him from here on.

‡ I said that the movie was funny, not that it was scientifically accurate.

philosophical—demeanor. I was about sixteen years old when I saw this movie for the first time, and while I loved it all, I found this part especially hilarious. I am most certainly not a teenager anymore, yet the memory of this scene still cracks me up.

A few decades after the release of *Caveman*, scientists found real-life, if circumstantial, evidence suggesting that dinosaurs most likely coexisted with psychoactive drug-producing organisms, raising the possibility that they might have crossed paths. This evidence appeared in a 2015 paper published in the journal *Palaeodiversity*, which described the discovery of some grass encased in amber[2] and dating back to almost one hundred million years ago, right in the middle of the Cretaceous period.[3] Finding a one-hundred-million-year-old plant inside some amber was not so extraordinary in itself; plants have been around for far longer than that, and we have many such specimens. The remarkable detail was that *this* grass specimen was infected with a type of fungus—and not just any fungus, but a species recognizable as belonging to the modern genus *Claviceps*, also known as *ergot*. These fungi are well known for producing (among other things) *ergotamine*, a medicinal compound with known psychoactive properties and a precursor to lysergic acid diethylamide (LSD), a well-known and very potent hallucinogen.

The authors of the *Palaeodiversity* paper speculated that certain species of herbivorous dinosaurs might have fed on these fungus-contaminated grasses and therefore could have experienced psychoactive effects. The paper references an earlier record that provides hard evidence that herbivorous dinosaurs did indeed feed on grasses, based on their presence in dinosaur coprolites (fossilized dung*).[4] It is possible, even likely, that the physiology and behavior of dinosaurs as well as their reactions to hallucinogens were similar to the physiology, induced behavior, and sensitivity to hallucinogens displayed by modern animals. Thus, it is not so far-fetched

---

* Who says science is not glamorous?

to imagine that, from time to time, a dinosaur might have eaten *Claviceps*-infected grass and suffered (or enjoyed) the usual consequences. As entertaining as this image is, I must reluctantly note that for a dinosaur to get high in this way, it seems probable that the behemoth would have to consume an awful lot (I mean, a *lot*) of psychoactive fungi–infected grass. Then again, we know absolutely nothing about the pharmacology of dinosaurs, so who can say? Perhaps dinosaurs were unusually sensitive to hallucinogenic drugs and just one small nibble of *Claviceps* would have done the trick.* It would be interesting to observe such effects in a dinosaur; alas, we will most likely never see an intoxicated dinosaur in real life. Happily for us, we do not have to rely so heavily on our imaginations when it comes to other kinds of animals—the animal kingdom is full of examples of creatures displaying intoxication-like behaviors when they consume certain plants and fungi. What's more, just as in the case of many humans, this consumption of psychoactive plants and fungi is often intentional, and has little to do with nutrition and everything to do with "self-medication," whether for health, recreation, or both.

---

There are three types of characters in virtually every story you will read in these pages, and when available and practical, I will include information about them all. The first is humans, from the ancient peoples who provide us with the earliest evidence of human chemical use, to the philosophers, scientists, and other thinkers whose curiosity led them to ponder and probe nature and later systematically develop our knowledge of drugs. Science is a human endeavor, and as such, human nature plays a central role in how science works and grows.

---

* In fact, over the years, there have been several satirical cartoons proposing in one way or another that the dinosaurs' demise was because of drugs, so this is hardly a new notion.

The individual personalities of the humans involved in research are invariably reflected in their discoveries, and that is part of their charm.

A second set of characters in these stories is the chemicals themselves. The only way to truly understand why a certain plant is psychoactive or medicinal is to have some understanding of the workings of the chemicals it produces. The characteristics of these chemicals often hold the key not only to how a substance affects various creatures, but also to the question of why a certain substance is produced in the first place. A confession: I love chemical structures, and I have included a variety of illustrations to share with you my fascination with their beauty.

The third, final, and perhaps most important cast of characters you will meet is made up of the nonhuman living organisms that make, obtain, use, and consume chemicals, whether for their advantage or leisure. These organisms include both members of the animal and plant kingdoms. The idea of animals engaged in active cooperation or competition with plants is not as familiar as their most direct relationship (namely, one eating the other), but it is at the center of the evolution of many psychoactive substances and their use. The animal characters we meet will range from the seemingly foreign to the actually domestic, but I will tell you in advance that the underlying assumption in this book is that even the strangest or smallest of our animal cousins is more like us than we might think.

While most of the stories we'll hear on our journey deal with animals (including humans) seeking or being exposed to psychoactive substances made by plants or microorganisms, it bears mentioning that our vegetal and microscopic friends do not have a monopoly on the production of psychoactive chemicals. Some animal species synthesize such compounds as well, and with a degree of chemical expertise that has nothing to envy in the synthetic abilities of plants and microbes. Moreover, some other members of the animal kingdom actively seek these animal chemists with the apparent purpose of experiencing the psychoactive states their chemicals induce.[5] Alas, although psychoactive animals are fascinating, space considerations preclude me from covering the topic here. This is a matter better left to a future book.*

---

In this book, our conversation will take place over the course of six chapters. The first four cover the scientific foundations that inform our journey, discussing some of the main players and most pressing questions and using examples from the plant, animal, and human kingdoms to illuminate these concepts. Do not fear, my intrepid reader! These chapters are full of scintillating stories (and hilarious history) that help put the three main characters we talked about above in their proper context. The last two chapters are more like a safari of sorts, two collections of tales (one covering invertebrates and one those with spines) of animals under the influence, whether naturally or through the auspices of experimentation. I will tell you a little more about what's to come in each chapter now, with the unapologetic hope of piquing your curiosity.

Chapter one begins with "liquid gold"—yes, good old alcohol, a substance that has undoubtedly shaped human history and, of course, remains popular in countless versions today. We will learn a bit more about alcohol itself, and encounter examples of animals imbibing, including some adorable little creatures with seriously massive tolerances. We'll also take a look at an ingenious contraption that measures drunkenness in fruit flies, and raise some questions, including that of how we came to consume psychoactive substances in the first place.

Chapter two gives us a brief grounding in pharmacological principles—including what constitutes a drug, and the fine line between medicine and poison. It also introduces the rest of our cast of psychoactive substances, and we'll hear about some of the brave

---

* Wink, wink.

and/or foolhardy scientists who used themselves as guinea pigs in studying these compounds (and the long naps and terrifying bicycle rides that ensued). Along with discovering the connection between a Jesuit priest, a barber, and cocaine—and exploring an out-there theory about the evolutionary significance of psychedelics—we'll also see how to keep your llama alert, and what some clever birds do with tobacco.

Chapter three explores the medicinal use of drugs, bringing us to the fascinating topic of animal consciousness—do animals "mean" to use these substances, and how can we know what they experience when they do? We'll travel deep into the mountains of ancient Iraq, where scientists discovered possibly the earliest evidence of the human use of medicinal substances. But while we humans have indeed been at this a long time, we'll see that our animal cousins got there first and (as in the case of a certain adorable porcupine) taught us by example. Prepare to be amazed by the sheer ingenuity of animals who use naturally occurring products to self-medicate.

Chapter four reveals the secrets of plants and fungi, who produce most of the staggering variety of psychoactive substances we find on this planet. They do not make these mind-bending chemicals as a gift for us; if you were surprised at how cleverly animals use psychoactive substances, wait until you read about how plants use them to compete in the game of life. We will untangle some remarkable relationships between plants and animals, and between insects and us, including a few fascinating hypotheses that try to explain why humans crave psychoactive substances.

Chapter five—part one of our closing safari—traverses the world of invertebrates small and large. We'll see what killer bees are like when they're drunk, and what happens to the webs spiders build under the influence of various drugs. We will drop in on the best-known insect alcohol enthusiast, the humble fruit fly, to see how alcohol affects their, well, "romantic" behaviors, and encounter possibly hallucinating sea slugs and the scientifically useful anxiety of worms. We'll end with what happens to octopi on ecstasy, and by the time the chapter is finished, I can guarantee that you'll see the spineless world in a whole new light.

Chapter six moves on to vertebrates, with tales of the drinking songs sung by zebra finches, elephants drunk on fruit and felled by LSD, and more. We'll cover both the well-documented and some interesting apocrypha (ahem, squirrels on speed), and encounter a variety of relationships between familiar animals and psychoactive compounds, including cats and catnip, horses and locoweed, and reindeer and . . . fungus-contaminated urine. I'll also take a moment to explain how goats (of all creatures) led us to the discovery of that most heavenly of beverages, coffee. The book closes with a look at one of the most intelligent mammals on Earth—dolphins, including the tantalizing possibility that various dolphin species engage in recreational intoxication using a substance that, as far as anyone knows, is not sought by any other animal for that purpose (porpoise?). We'll also see how a scientist who gave LSD to dolphins sold the idea of studying dolphin communication to astronomers hoping to search for extraterrestrial intelligence. Science is full of these unexpected connections—and unexpected connections are at the very heart of this book.

So let's get to it.

CHAPTER 1

# GETTING INTO THE SPIRIT OF THINGS

*All animals are charmed by opium. Addicts in the colonies
know the danger of this bait for wild beasts and reptiles.
Flies gather round the tray and dream, the lizards with
their little mittens swoon on the ceiling above the lamp and
wait for the night, mice come close and nibble the dross . . .
The cockroaches and spiders form a circle in ecstasy.*
—JEAN COCTEAU, *OPIUM: THE DIARY OF HIS CURE*

*I'm no wildlife expert, but I think this monkey might be drunk.*
—ANONYMOUS MEME\*

To the question "Do animals have minds like ours?" the short
answer, alas, is, "Who knows?" For one thing, it depends upon
what you mean by "like ours," but even when we get more specific,

---

\* This is the caption of a very funny meme that shows an obviously
photoshopped monkey wielding a stick and sneaking up behind a magnificent
specimen of the king of the jungle. The monkey's obvious intention is to
strike the lion with said stick; it is not hard to imagine the likely evolutionary
outcome of this act. I have tried (unsuccessfully) to trace the meme's origin; if
you know it, please enlighten me so I can give proper credit where it is due.

9

various caveats leave us with shades of "maybe" (though if you ask me, I vote for "likely"). What we *can* say is that many behaviors we think of as exclusive to humans are in fact also expressed by our animal cousins. In these pages, we will explore one particular category of behaviors that animals share with us (or us with them): the consumption of psychoactive substances. That animals engage in this practice is no secret; ancient societies have known and commented on this for centuries, perhaps millennia. For those of us in modern societies who live in predominantly urban areas, however, it may come as a surprise to hear just how far beyond your feline companion's love for catnip this tendency extends. Take, for example, that lovable mascot of the land down under: the koala.

The koala (*Phascolarctos cinereus*) is a marsupial bearing a general resemblance to a small bear, and in fact, *Phascolarctos cinereus* literally means "ash-colored pouched bear."[1] Koalas are undeniably endearing and generally pleasant, but they are wild animals nonetheless, and there are documented cases of koalas attacking people and their pets; these furry little bears may be nice, but they are not pushovers. (They are also an endangered species, a situation that has become even worse due to the forest fires that devastated large areas of Australia as I was writing these words.) Koalas feed exclusively on several species of eucalyptus trees, a risky choice of meal made possible by the unique microbiome of the koala's digestive system—eucalyptus plants produce a plethora of toxic chemicals that the plant uses for protection against its predators. Of course, as with all mammals (and especially marsupials, which are born in an undeveloped state), a baby koala's first food is its mother's milk. As a baby koala grows, it graduates to solids: specifically, its mother's feces (a practice present in other mammalian species as well, though happily not our own). Eating mom's poop helps the babies acquire the various bacterial species that will eventually aid them in digesting an all-eucalyptus menu.

Some accounts describe the smell of koalas as reminiscent of cough drops, for obvious reasons. Given the opportunity, however, it is entirely possible that koalas would smell instead like a cross between Captain Morgan and the Marlboro Man. You see, as it happens, the cute and cuddly koala has a taste for alcohol and tobacco.

I first learned about this through the work of Charles Darwin (a man who surely needs no introduction). In writing his books, Darwin often relied on information and specimens sent to him by various pen pals over the years—a prolific writer, he kept up an impressive amount of correspondence with many like-minded people. Remarkably, he was able to do so solely through what we know today as "snail mail" and without even the option of a telephone call. I often wonder what he might have done had he lived in our era of social media and instant communication. I think that at the very least he would have been a popular blogger, perhaps a podcaster, even.[2]

One of Darwin's correspondents was Robert Arthur Nicols, a Fellow of the Royal Geographical Society. He wrote at least eighteen letters to Darwin, sometimes with questions, sometimes with information he thought Darwin might find useful. In a letter dated March 7, 1871, Nicols wrote about our friend the koala:

> *While in the Colony of Queensland some years ago, I kept as pets, at different times, three individuals of Phascolarctus [sic] cinereus, each of which manifested an inordinate love of tobacco and rum. I never in any case invited the desire: it appeared always to originate in their sense of smell. Thus, sitting on my shoulder, his usual place in the evening, Phascolarctus [sic] would reach forward and clutch my pipe, sucking the saturated wood or clay with avidity, and licking even the bowl regardless of heat: when denied this luxury he would sit and inhale the smoke with closed eyes, licking his lips the while, and appearing to be in a state of dreamy enjoyment. When the taste was once established he would seek out and chew every pipe he could find, thus imbibing an amount of oil, which, I confess, would have killed me. The three cases were substantially the same. And I have heard on reliable authority, of others.*

Nicols continued:

*It was the custom to have a glass of rum in the evening and each of these animals, by first licking the spoon probably, got a taste for the liquor after which it was impossible to prevent them from seizing upon any utensil containing it, and finally they were allotted a small quantity, neat as they preferred it, when they would retire to the rafters and sleep. Occasionally they would become excited and bite severely, their usual disposition being extremely mild, when under the influence of the spirit.[3]*

The koala is not alone, and Darwin documented similar instances in several of his books. In the first few pages of *The Descent of Man* (the main message of which is similar to that of this book—namely that animals are not so different from us), he discusses disease states that occur in both animals and people; he follows his thoughts on disease by making a point about medicines, namely that medicines *"produced the same effect on them [animals] as on us."*[4] To further get this point across, he gives a few examples of animals that took, well . . . *not* medicines:

*Many kinds of monkeys have a strong taste for tea, coffee, and spirituous liquors: they will also, as I have myself seen, smoke tobacco with pleasure. Brehm[5] asserts that the natives of north-eastern Africa catch the wild baboons by exposing vessels with strong beer, by which they are made drunk. He has seen some of these animals, which he kept in confinement, in this state; and he gives a laughable account of their behaviour and strange grimaces. On the following morning they were very cross and dismal; they held their aching heads with both hands and wore a most pitiable expression: when beer or wine was offered them, they turned away with disgust, but relished the juice of lemons. An American monkey, an Ateles [a spider monkey], after getting drunk on brandy, would never touch it again, and thus was wiser than many men. These trifling facts prove how similar the nerves of taste must be in monkeys and man, and how similarly their whole nervous system is affected.[6]*

## THE SUBSTANCE THAT STARTED IT ALL

It seems only right that we begin our journey with the molecule that very likely introduced humans to the phenomenon of the psychoactive: this chemical is, of course, alcohol—specifically ethyl alcohol, or simply ethanol (Figure 1.1).* An ancient companion of humanity, I think it likely that no other chemical has contributed more to our collective history. Alcoholic beverages were and still are an integral part of a variety of cultural practices; they're used in social, ceremonial, celebratory, and religious rituals, or simply to keep us warm on cold winter nights. Unsurprisingly, the manufacture of alcohol is one of the largest (and oldest) industries worldwide.

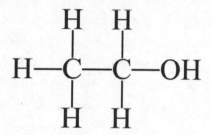

Figure 1.1. Ethanol. Drawn by the author.

Despite its influence and importance, ethanol is a simple substance, naturally produced by a diverse group of organisms. Curiously, however, the best available evidence tells us that its formation and/or synthesis do not necessarily require any kind of biological process. There are many examples of chemical reactions having nothing to do with living organisms that are capable of producing ethanol. Ethanol is also curious because, in contrast with other psychoactive compounds (the usual suspects: nicotine, morphine, cocaine, etc.), ethanol

---

* Unless I specify otherwise, I will use the terms "alcohol," "ethyl alcohol," and "ethanol" interchangeably.

does not seem to have a well-defined receptor target* that accounts for its biological properties. Yet alcohol is undeniably a psychoactive substance—there is good reason that ethanol has been called the "universal intoxicant."[7]

## IT CAME FROM OUTER SPACE

It is not hard to find ethanol in nature; moreover, as alluded to previously, ethanol can be of nonbiological origin. In fact, ethanol is so common that there are significant amounts—and by that I mean a *lot*—in, of all places, interstellar space. This is hardly newsworthy; astronomers have known for about fifty years that there is ethanol in space. In 1975, a study of the astronomical object Sagittarius B2 (Sgr B2—essentially a huge space cloud) determined that it would yield approximately 7,500,000,000,000,000 cubic kilometers of pure ethanol. This is much more than the total amount of ethanol produced by us throughout history so far (and likely ever).[8] To get some additional perspective on just how much ethanol there is in Sgr B2, the volume of all the water within our seven seas is only about 1.4 billion cubic kilometers, give or take, meaning that the alcohol in Sgr B2 would fill about 53 million Earth-size oceans.

That's a lot of margaritas.

Ample evidence indicates that humans have had a close relationship with alcoholic beverages for at least the last hundred thousand years (and in all likelihood much longer than that), but the available

---

* Briefly, receptors are proteins in cells that react to the presence of a chemical. We will discuss this term further in the "Philosophy of Pharmacology" section in chapter two.

historical record of this relationship is no more than ten thousand years old.[9] While the conventional wisdom about humanity's decision to get into the grain business has generally held that we did it for food, in a paper published in the 1950s, an intriguing idea was proposed by Dr. Robert J. Braidwood, then an archaeologist at the University of Chicago. His paper—charmingly titled "Did Man Once Live by Beer Alone?"—proposed that the original motivation behind our cultivation of cereal-producing crops was not exactly nourishment related. In a nutshell, Dr. Braidwood's thesis was that while our ancestors did indeed eventually select agricultural crops for nutritional purposes, they first selected and experimented with particular crops not for their ability to produce food, but for their ability to produce fermented beverages (specifically beer).[10] My own guess—admittedly just that—is that the truth about what nudged us to practice agriculture is somewhere between food and booze, as it were. Alas, this development predates written records, and it is very hard to figure out which came first by other means; this is yet another aspect of our collective history that will likely remain a mystery. One thing is certain, however: nowadays, as before, we cultivate grains for both nutrition and for use in alcoholic preparations.[11]

Interestingly, it seems that at least part of the motivation our forebears had for preparing alcoholic solutions and beverages in the first place went beyond the desire to experience their psychoactive effects, showing the chemical versatility of ethanol. Starting around the thirteenth century, early pharmacists began preparing solutions containing various concentrations of ethanol, ostensibly for medicinal purposes. Depending on the specific alcohol strength, these liquids were referred to as *aqua ardens* (burning water), *aqua flamens* (flaming water), or *aqua vitae* (water of life).[12] This last moniker made a lot of sense for people living in those times, since contaminated water was a significant health hazard. This reality was especially relevant in light of our medieval ancestors' lack of knowledge about microorganisms and their role in disease. They may not have understood why, but people did not fail to notice that, all else being equal, they rarely got sick when drinking "spiked" beverages versus water from the local well. Our forebears also noticed that rubbing alcohol–rich solutions on damaged skin prevented and even cured infections. In time, alcohol became one of the main ingredients of multiple remedies. (I strongly suspect that the accompanying "buzz" must not have been terribly unwelcome.)

But grain on its own does not contain ethanol. And while, as described previously, ethanol is abundant in the universe, we cannot—yet—"go get it" from outer space. So where does the alcohol that humans use come from? As it happens, we've long relied on the help of microscopic friends in this matter, by obtaining ethanol courtesy of the humble yeast.[13]

## The Yeast Among Us

Some of the best-known alcohol-producing organisms are a series of yeast species belonging to three main genera: *Saccharomyces*, *Dekkera*, and *Schizosaccharomyces*; probably the best-known member of the bunch is the common baker's (and brewer's) yeast, *Saccharomyces cerevisiae*.\* These three yeast lineages have been evolving away from each other for at least two hundred million years, yet they've all retained the capacity to engage in alcoholic fermentation, suggesting these biochemical pathways are not only ancient in nature, but useful in some way.[14] When a particular biochemical trick is conserved widely, it is bound to be important.†

So perhaps you are wondering: *What's in it for the yeast?* In other words, from the yeast's perspective, why make ethanol at all? This is an important question because ethanol exposure is potentially dangerous to yeast; let's not forget that alcohol is a well-known

---

\* Yeasts are nothing more than unicellular fungi, but *S. cerevisiae* is an extensively studied model organism in the biomedical sciences.

† For some additional perspective on that idea, see "An Ancient Partnership" on page 18.

disinfectant: disinfectants kill microorganisms, and yeasts are, well, microorganisms.

Ethanol kills microorganisms via osmotic stress (briefly, disturbances in water transport), the generation of noxious forms of oxygen (yes, oxygen can be toxic),[15] and the disruption of membrane structures, among other quite nasty effects, nearly all of them lethal to most microbes. As you can imagine, this is why we use alcohol against microbes in the first place. So, again, why would yeasts make ethanol? Part of the answer to the yeast/ethanol riddle is that many species use ethanol as a source of energy; it turns out that many yeast species are no exception. However, it is unlikely that this was the original purpose of the evolution of ethanol production in these organisms, precisely because of its aforementioned toxicity. No, the key piece of information was the discovery that *S. cerevisiae* and related species have evolved unusual resistance to relatively high ethanol concentrations. Thus, the current understanding of the yeast-ethanol relationship is that yeasts produce ethanol to use for fuel, but also as a defensive strategy against other microbial competitors. In other words, yeasts produce ethanol at concentrations that will kill off the competition but that are not high enough to harm themselves. The unusual resistance of yeasts to alcohol is due largely to the molecular machinery inside their cells that gives them the ability to detoxify ethanol by transforming it into less toxic chemical species.[16] Thus, using a neat series of biochemical tricks, yeasts are able to generate energy from alcohol while also using it as a chemical defense. Many naturally occurring compounds produced by living organisms were evolved to be used by them as chemical weapons: alcohol is no exception. (We will explore this concept in more detail in chapter four.) More likely than not, the bioenergetic advantage of ethanol resistance was a happy evolutionary accident that was dutifully picked up and taken advantage of by natural selection.

## AN ANCIENT PARTNERSHIP

Biological life as a phenomenon has three nonnegotiable imperatives—information an organism requires in order to survive: *what to eat, what to avoid, and what to mate with.* I stated it this way in my 2018 book, *Strange Survivors*, but I am by no means the first to articulate this idea; some version of it is at the heart of all biology, and it is perhaps the closest we get to a universal "biological law." The first piece of the puzzle (what to eat) refers to the acquisition of energy from the environment. The very idea of metabolism (the summary of all the chemical reactions in an organism) pretty much boils down to acquiring food from one's surroundings and trans- forming this food into chemically useful energetic molecules, predominantly *adenosine triphosphate* (ATP), which is the *universal* chemical fuel for all biological life on this planet. A few paragraphs back, we mentioned that ethanol, albeit potentially toxic, is also an energy-rich molecule. As a result, during the course of evolution, many different yet related metabolic pathways have strived to squeeze energy from ethanol to make a bit of precious ATP. The main enzymes at work in virtually all organisms capable of metabolizing—that is, consuming and making use of—ethanol are *alcohol dehy- drogenase* (ADH) and *acetaldehyde dehydrogenase* (ALDH). These two enzymes work in tandem (for a reason we will see in a moment); there are versions of this team in every known species on Earth, from bacteria to blue whales. Further, these two enzymes have been working—and evolving—together for a long time.[17] Depending on the specific method of analysis, evidence suggests that the ADH/ALDH pair has collabo- rated for the better part of the last three billion years.[18] In humans alone, there are between twelve and seventeen sets of genes that codify these enzymes, each one with various alleles (alternate gene forms).[19] And as we've alluded to once already, if a system is ancient and conserved throughout evolutionary time frames (millions of years or even longer), this system is bound to be important; there is no plausible alternate interpretation.

The reason organisms need both ADH and ALDH working together in order to effectively metabolize ethanol is quite interesting. Technically speaking, it is ADH, not ALDH, that metabolizes ethanol. So ethanol comes in and ADH goes to work. The main product of the reaction catalyzed by ADH is a molecule called *acetaldehyde*, which is a problematic molecule in itself because at high enough concentrations it becomes toxic, causing genetic mutations and wreaking miscellaneous havoc. For example, acetaldehyde is one of the main factors in the liver damage that can occur from excessive alcohol consumption. Admittedly, at first glance, converting one toxic molecule (ethanol) into another toxic molecule (acetaldehyde) does not seem like much of an improvement. Happily, this is when ALDH swoops in and comes to the rescue. ALDH metabolizes acetaldehyde into *acetic acid*, a (pretty much) harmless molecule that can go straight into ATP-producing pathways, albeit a tad less efficiently. In short, the ADH/ALDH duo not only detoxifies ethanol and acetaldehyde, but also uses them as precursors to make ATP; in this proverbial one-two punch, everybody wins. Not surprisingly, the function (or dysfunction) of the ADH/ALDH gene pair affects virtually every aspect of ethanol metabolism in nature. Moreover, it seems that the evolution of fruit-eating behavior (fermented fruit being a common source of naturally occurring alcohol) closely followed the evolution of the activity of the ADH/ALDH gene pair; as this optimized the activity of these enzymes, it resulted in better handling of ethanol at the organism level (that is, the development of higher alcohol tolerance and so on).

Biochemistry may not seem like a topic appropriate to a night at your favorite watering hole, but next time you're out, try sidling up to the bar and asking a fellow patron: "Metabolize here often?" You just might make a new friend!

## When Your Favorite Brewery Is Within You

An unusual medical condition results in some humans appearing to produce ethanol on their own. It is characterized by the appearance of ethanol in a person without any apparent alcohol ingestion, resulting in the well-known symptoms of ethanol intoxication (namely, apparent drunkenness).[20] This condition goes by various names, including *gut fermentation syndrome*, *endogenous ethanol syndrome*, and *drunkenness disease*, but the most common—and my favorite—is the apt *auto-brewery syndrome* (ABS).

Although it stands to reason that this must be an ancient disease, the earliest circumstantial reporting of an ABS-like case in medical literature is generally said to date from 1948, in a paper describing the sad story of a five-year-old child who died from a ruptured stomach. In this case, the autopsy indicated an abnormal amount of gas in the GI tract as well as a pervasive ethanol smell; however, blood ethanol was not measured. The physicians in charge speculated that ethanol had been produced in the boy's GI tract via bacterial fermentation of sweet potatoes the child had ingested. The first formal medical report that described ABS dates from 1952 and details the case of a forty-six-year-old man who became inexplicably intoxicated after exploratory surgery. The attending physicians identified *Candida sp.* (yeast) as the likely origin of the ethanol causing the patient's intoxication.[21]

A set of collected cases describing ABS was published in 1972, and from then on the syndrome became slightly better known, although to date it is still considered a widely underdiagnosed condition.[22] The degree of intoxication that results can vary widely based on a number of factors, as you might imagine. But there is no doubt that there have been a significant number of people throughout history mistakenly perceived as being "drunk" because of this syndrome.

We've developed a more complete historical perspective of ABS only recently, thanks to Dr. Barbara Cordell, an expert who has published academic papers on ABS and, in 2019, a popular science book on the topic, *My Gut Makes Alcohol!: The Science and Stories of Auto-Brewery Syndrome.* Dr. Cordell notes that research suggests the phenomenon of "gut fermentation" was somewhat in vogue among the physician set in the 1930s and '40s, and while the 1948 case is often referenced as the first in the literature, a 1913 article by a Dr. Robert Saundby in the *Proceedings of the Royal Society of Medicine* mentions "auto-intoxication" in describing an eighteen-year-old boy who presented as intoxicated due to intestinal fermentation, which then resulted in acetone poisoning. Says Dr. Cordell:

> [I]n fact, Charles Bouchard, a professor of Pathology and Therapeutics in Paris, authored a 368-page tome in 1894 titled "Auto-Intoxication in Disease," that was translated to English in 1906. In short, it appears science had well-accepted Auto-Brewery in the early 20th Century.

We humans, of course, lack the proper biochemical pathways to produce ethanol—most of us can eat all the sweet potatoes we like without fear of them fermenting unexpectedly. In patients with auto-brewery syndrome, the likely culprit is just what the physicians in the 1952 case suspected: yeast. In most patients the fermentation comes courtesy of *Saccharomyces cerevisiae*, a common yeast that naturally colonizes our bodies (though other microorganisms can also contribute to the development of this disease, including *Candida albicans, Candida krusei, Candida glabrata*, and even a non-yeast, *Klebsiella pneumoniae*). The available evidence seems to indicate that individuals affected by this syndrome tend to be immunocompromised and therefore less able to withstand and/or eliminate yeast overgrowth. In these individuals, ABS may flare up when consuming high amounts of carbohydrates. Studies indicate that not just nutritional habits but antibiotic use and general health are linked to ABS.[23] In general, this syndrome is responsive to fungicides and diet changes, but some sufferers have multiple relapses and the condition can become chronic.

In affected patients, the concentration of ethanol in the blood gets close to (or often exceeds) the legal limit for a person to be considered intoxicated, and so you may not be surprised to hear that people have attempted to use auto-brewery syndrome as a legal defense to get out of DUI charges. These attempts have usually been unsuccessful, however in a recent case a woman was indeed able to satisfactorily demonstrate that her high blood-alcohol level (reportedly four times the US legal limit) was due to ABS, and hers is unlikely to be the only such instance.[24]

The condition has gained popular notice as it has begun to pop up on medical shows like *Grey's Anatomy*, and variations of this syndrome are appearing in the popular as well as the medical literature. One recent case tells the tale of a woman with a rare version of ABS in which she produced ethanol in her bladder, leading to the detection of alcohol in her urine.[25] In some patients, other diseases like diabetes and Crohn's disease can coincide with auto-brewery syndrome, but this is not always the case, and the connection is still being investigated.[26] Sadly, it seems that auto-brewery syndrome may even rarely contribute to cases of sudden infant death syndrome.[27] There is much we still don't know, but I am sure that as these words are being written, there are biomedical scientists furiously researching this disease in all its variants.

ABS requires the work of unicellular tagalongs, like yeast; virtually no multicellular organisms can naturally produce ethanol. The exceptions are rare: certain plants and (even more rarely) some species of fish, and in the case of the fish, this ability usually only presents in very cold environments. The hypothesis that certain fish might produce metabolic ethanol in response to freezing temperatures in their environment has been around since the late 1980s, based on some odd metabolites noticed by researchers.[28] However, it was less than three years ago that it was discovered that at least two species of fish (oddly, one of these is the common goldfish, a fish that certainly does not hail from the Arctic) do indeed synthesize ethanol under these extreme conditions.[29] But nature is prolific; I suspect we are bound to discover many other ethanol-producing animal species sooner rather than later.

## TOUGH LITTLE DRINKERS

One of the "themes" of this book is that evolution is economical; it tends to keep what works reasonably well and avoid reinventing the wheel. Prime examples of this are certain species of mammals who count flower nectar as part of their diet. Scientists have long speculated that mammals who feed on nectar are bound to cross paths not infrequently with fermented nectar, with the obvious pharmacological consequences. And, as it happens, a few of the mammals who regularly pay visits to flowers with fermented nectar display an unusual characteristic. They are unusual mammals to begin with, mind you. Two of them are primates, though not typical primates at that; the third one is "kind of" a primate. The two primates in this story are the slow loris (*Nycticebus coucang*), which hails from Southeast Asia and nearby archipelagos, and the aye-aye (*Daubentonia madagascariensis*), a type of lemur from Madagascar. The third mammal—the "kind of" primate—is the pen-tailed tree shrew (*Ptilocercus lowii*); these little fellows live in regions of Malaysia, Borneo, and nearby islands. Do not tell my zoologist friends, but it does not matter to us whether these fascinating organisms are primates, kind of or otherwise; what interests us is their common claim to fame. We'll get to that in a minute, once they have been properly introduced.

The slow loris is exceptionally cute, even for a mammal—among the cutest, depending on who you ask. Small and sweet looking, slow lorises are between eight and twelve inches long, with big eyes that make them appear worried or sad. One of their signature features is their tongue: one of the longest (proportionally) in the animal kingdom, they use it to get at the nectar in flowers. An excellent reminder of the inadvisability of judging a book by its cover, these adorable creatures are also poisonous (in fact, lorises as a group are the only poisonous primates known so far), and vicious to the point that they must be kept separated when in captivity; otherwise, they will literally kill each other.*

The aye-aye is, well . . . not as cute as the slow loris. In fact, it is difficult to imagine anyone (with the exception of another aye-aye) describing them as cute at all. To get a general idea, picture a long-tailed scruffy squirrel with outsize ears, a prominent nose, and perpetually round, wild-looking "crazy dude" eyes. The best description of the aye-aye I have heard—bar none—comes from my friend Mr. Micah Hanks, an accomplished journalist, writer, and podcaster, who remarked that the aye-aye "looks more like a chupacabra than like a lemur."[30] Alas, the aye-aye *is* a type of lemur, one with an interesting anatomical feature: a very long middle finger on each hand. (Admittedly, this does nothing to increase their visual appeal.) Aye-ayes use this long, thin finger to tap on wood to locate the hiding places of wood-boring insect larvae, and then to "fish" the grubs out from their nest, consuming the nutritious snack with gusto—they

---

* Slow lorises are fascinating animals, and I talk more about them in *Strange Survivors* (pages 108 to 110), should this introduction have piqued your interest.

also use this finger much as the slow loris does their tongue, to fish out nectar. Aye-ayes are the largest native primate in their stomping grounds of Madagascar and the largest nocturnal primate in the world, though they clock in at about twelve inches, tops. Although they possess a timid temperament consistent with their solitary lifestyle, when they encounter each other in a "non-procreative" context, they can get aggressive.

Our third creature of interest is the pen-tailed tree shrew: about five inches long, these wee creatures get their name from their beautiful tails, which look very much like the feather of a quill pen. Not your typical shrew, this species diverged from the rest of the tree shrews some twenty million years ago. The pen-tailed tree shrew is especially interesting to evolutionary biologists as a hypothetical model for an "ancestral primate," in part because it has evolved little in the last twenty to thirty million years.[31] On the other hand, many in the scientific community contend that pen-tailed tree shrews are not true primates, since they are descended from another mammalian branch (but remember, just between us, this does not really matter for the purposes of our chat).

Aside from their mammalian nature, at first glance you might not think that slow lorises, aye-ayes, and pen-tailed shrews have very much in common, and you would be—mostly—right. However, they share an interesting behavioral "superpower": their ability to tolerate significantly higher amounts of alcohol than other mammals. Scientists have observed these creatures feeding on flowers with fermented nectar without apparent intoxication—for example, they remain equally agile and alert before and after their nectar consumption. The secret behind their alcohol tolerance is still not completely clear, but it likely owes a debt to some unusual biochemistry.

Take the recent research of Drs. Samuel R. Gochman, Michael B. Brown, and Nathaniel J. Dominy of Dartmouth College in New Hampshire, scientists who became curious about a certain genetic oddity—a mutation in a version of one of the enzymes we met recently: alcohol dehydrogenase (ADH)—that is in charge of metabolizing ethanol in most organisms. This mutation is present in the great apes (including us) and increases the efficiency of ethanol processing in that specific ADH type by about fortyfold;* it was also discovered to be present in another, evolutionary distant primate . . . the aye-aye. This got Gochman and his fellow scientists at Dartmouth thinking. Organisms possessing this or similar variants might be capable of better metabolizing ethanol and consequently better able to take advantage of its high caloric content while minimizing its harmful effects, including intoxication. Based on this reasoning, such organisms would tend to feed on fermented fruit (and nectar) with few or no consequences. Aye-ayes, along with fellow primates the slow lorises, were speculated to consume "non-trivial" amounts of fermented nectar as part of their diets, and were at least known to come in contact with the stuff. Could this constellation of information be used to better understand our fermentation-friendly pals?

Admittedly, Gochman, Brown, and Dominy had very little data to start with. For example, there was no information about the ADH varieties present in slow lorises compared to aye-ayes, nor specifically whether slow lorises have the same ADH mutation present in aye-ayes and the great apes. Moreover, there was no hard data on whether either organism actually *preferred* fermented nectar over "regular" nectar (to date there is very little data available on this matter, especially in wild populations). However, this situation represented precisely the kind of challenge a scientist craves: having very little data, a speculation-rich topic, and ideas amenable to experimental observation. They designed what we would call a pilot experiment, aimed at testing whether either aye-ayes or slow

---

* It is important to keep in mind that these are generalities. Alcohol intoxication has a lot to do with the weight of the "drinker," and, as is the case with most genes, alcohol-related genes display variability across a population. Case in point: we all know people who can hold their liquor, and others who get dizzy with a single beer (like yours truly).

lorises prefer sugared solutions containing ethanol at concentrations between 4 and 5 percent.[32] In true scientific form, they worked with what they had available: two aye-aye specimens and one slow loris. The two aye-ayes were named "Merlin" and "Morticia"—extremely apt aye-aye names, if you ask me—while the slow loris went by the misleadingly peaceful-sounding "Dharma."*

One would think that the limited number of experimental subjects alone would weaken the statistical strength of the data, and yet they obtained very compelling results. Not only did all three animals willingly feed on alcohol-spiked sugar water, they clearly preferred it over an equally sweet nonalcoholic drink. Furthermore, given the choice of two drinks with different alcohol concentrations, they always preferred the "spikier" one. What's more, they did not seem to display any inebriation-related behaviors. The authors cautiously interpreted their data as consistent with Dr. Robert Dudley's oft-referenced "drunken monkey hypothesis,"† but correctly admitted that much work still needs to be done to understand the mechanisms behind this apparent preference and tolerance for alcohol, especially in light of the various enzymes that participate in ethanol metabolism.

As for the third member of our original trio of hollow-legged organisms, the pen-tailed tree shrew's ability to imbibe alcohol-laced nectar at concentrations that would make most other mammals swoon has also been the subject of scientific investigation.[33] At Universität Bayreuth in Germany, Dr. Frank Wiens and his collaborators tested the alcohol tolerance of pen-tailed tree shrews and a few other small mammals by measuring the amount of a specific product of

---

* Their names may not be scientifically significant, but having learned them, I could hardly keep them from you.

† The drunken monkey hypothesis posits that animals acquired selective advantage by having the ability to forage for the most nutritious, ripe fruit via their natural attraction to the smell and taste of alcohol. We will talk more about the drunken monkey hypothesis in a later chapter.

ethanol metabolism (ethyl glucuronide, EtG for short) in the hair of these little fellows. The idea behind this test is that the higher the amount of EtG found in a hair sample, the more alcohol the hair's owner has been exposed to (presumably by indulging in fermented nectar). Pen-tailed tree shrews showed significantly higher levels of EtG compared to control animals like small monkeys and common rats, supporting the notion that these shrews not only encounter and consume alcohol in their natural habitat, but also tolerate it well. A surprising finding of this research is that two other creatures who live in the same neighborhood as pen-tailed tree shrews, namely the common tree shrew (*Tupaia glis*)* and the plantain squirrel (*Callosciurus notatus*), were found to have equally high levels of EtG in their hair, suggesting that there may be a variety of adorable little drinkers in the Malaysian forests. There is little doubt that further studies of these remarkable organisms will contribute to our understanding of the effects of alcohol in mammals—us included.[34]

## LORE OF THE FLIES

In the title of this book, I promised you drunk flies. We'll revisit our friend the fly many times throughout our conversation on animals and intoxication, as there is quite a lot to say about this tiny creature, but this first chapter on the first drug is a good place to start.

Otherwise known as the common fruit fly, *Drosophila melanogaster*† was the indisputable hero of twentieth-century genetics and developmental biology research and is still a significant source of

---

* Though a different sort of beast, the pen-tailed shrew is also a tree shrew because of where it makes its home.

† As of 2018, the genus *Drosophila* includes some 1,600 different species, although as with many other types of organisms that actual number is bound to be higher on account of undiscovered species. We'll mostly limit ourselves to *Drosophila melanogaster*.

valuable scientific information and insight today. Not to diminish *Drosophila*'s accomplishments in service of science, but it must be said that serendipity played a central role on the "rise of the fly" as an experimental organism. Thomas Hunt Morgan (arguably the driving force of genetics research in the late nineteenth century and on into the twentieth) essentially narrowed his search for a suitable animal model for genetics study to the fruit fly and an unapologetic favorite of mine: the planarian. Both fly and flatworm were excellent candidates, but for various reasons mostly to do with the specific type of study Morgan favored, the fly "won," and the larger research community followed his lead. Partially as a result, the humble fruit fly is perhaps the best biologically characterized and understood multicellular organism to date. A thorough exploration of *Drosophila*'s biology would fill another book entirely and at any rate is brilliantly explored elsewhere;[35] the main point for our purposes is that, because of the abundant and readily available biological information on *Drosophila*, from molecular goings-on to behavior, it was only a matter of time before this unassuming insect caught the proverbial eye of a small army of scientists interested in the phenomenon of drug consumption and addiction.

Happily, the fly did not disappoint, and *Drosophila* has turned out to be an excellent animal model for drug-abuse research. For example, as documented in thousands of published papers, scientific studies demonstrate that *D. melanogaster* reacts in the "expected" manner to exposure to pretty much all "traditionally" addictive substances.* Furthermore, drug-induced behaviors in the fruit fly closely parallel behavioral responses in vertebrates. To give this fact some perspective, let's not forget that, in an evolutionary sense, the insect and mammal lineages are separated by some seven hundred million years or so![36] And yet, *Drosophila* displays many addiction-related behaviors that

---

* In fact, as of January 2020, a simple Pubmed search (ncbi.nlm.nih.gov/pubmed) combining "*Drosophila*" and "drug" as keywords returns more than ten thousand published papers.

we usually associate with humans, among them habituation, tolerance, and reward. What's more, in *Drosophila*, just as in vertebrates, the neurotransmitter dopamine seems to control these effects.[37] As you may imagine, then, one of the addictive substances to which our friend the fly is sensitive is good old ethanol.

Paradoxically, despite the fact that some of the other names for *Drosophila* include the aforementioned *fruit flies*, as well as *pomace flies* and *vinegar flies* (these last two monikers intended to signify that they are often to be found around decaying and thus likely fermenting fruit),[38] concrete evidence indicating that *Drosophila* actually "liked" ethanol was lacking until relatively recently. To the best of my knowledge, the first study of alcohol dehydrogenase (the main enzyme that degrades ethanol) in *Drosophila* appeared in print more than fifty years ago,[39] but formal as well as anecdotal evidence indicating a strong link between fruit flies and alcoholic beverages predates these biochemical studies by a very long time. Moreover, *Drosophila* is hardly the only insect attracted to the fruit of fermentation. It is common practice to catch butterflies and moths by setting a beer or wine trap. In essence, the flying creatures detect the smell of the alcohol-laced liquid, fly toward it, and are trapped for collection, research, or pest-control purposes.[40]

Interestingly, if not totally unexpectedly, it seems that in *Drosophila* the genes that control food intake also control alcohol intake (which may not be a coincidence).[41] And as expected from an organism colloquially named for its preferred alcoholic food source, *Drosophila* possesses a certain resistance to the toxic effects of ethanol, just like our old friends the yeasts, as well as the curious primates that we saw earlier. At the molecular level, their resistance comes courtesy of the familiar team of ADH/ALDH.[42] This tolerance varies from fly to fly—and here I imagine I see you holding up a hand to stop me. *Wait a minute*, you say. *How can we tell whether a fly is drunk in the first place?*

I am so glad you asked.

## The Inebriometer

By necessity, telling whether a fly is drunk involves observing its behavior, as there is no way for us to communicate with a fly (and even if there were, the fly might lie and insist it was perfectly sober). We can infer drunkenness in a fly by straightforward qualitative observation—for example, "That fly looks sluggish and is flying erratically!" To *quantify* its drunkenness in true scientific fashion, however, you might wish to use a specialized instrument. Happily, there is in fact such an instrument available; it is called the *inebriometer* (Figure 1.2).*

Interestingly, this contraption was not created to explore drug-related phenomena; rather, it was invented as a tool to explore unrelated questions of evolutionary biology. The scientific world heard about the inebriometer for the very first time some thirty-five years ago, in a 1985 paper titled "Latitudinal Cline in *Drosophila melanogaster* for Knockdown Resistance to Ethanol Fumes and for Rates of Response to Selection for Further Resistance," authored by Drs. F. M. Cohan and J. D. Graf. Three years later, Dr. Kenneth Weber's doctoral dissertation from Harvard ("The Effect of Population Size on Response to Selection") described the concept development, construction, and use of a "Mark II" version of the inebriometer, with a few bells and whistles added.

Drs. Cohan, Graf, and Weber designed and optimized the inebriometer to classify individuals in a population of flies into several groups, namely: (1) the ones that got drunk very quickly, (2) the ones that took longer to get drunk, and (3) the ones that needed a really long time (or *a lot* of alcohol) to get drunk (or that never got drunk

---

* Yes, this is the actual name of this scientific instrument! However, Dr. Vincent Dethier, whom we will meet in a later chapter, coined the name of a similar instrument, the *drinkometer*, in his 1962 book, *To Know a Fly*—a delightful exploration of science and its methods using various types of flies as characters to tell the story. Incidentally, later we will talk about a modification of Dr. Dethier's drinkometer used to study alcohol responses in honeybees (Ford and collaborators, 2004).

at all). The intended use of the inebriometer was simply to separate a population of a particular species of flies into several subpopulations based on alternate phenotypes (expressions of genetic characteristics), these phenotypes being the various degrees of tolerance to alcohol intoxication expressed in the individuals of said fly population. Once individual flies of different phenotypes had been identified based on their resistance to alcohol fumes, studies could be conducted on the evolution of the genetic traits responsible for alcohol metabolism in the fly.

Of course, to accomplish this, these scientists first had to tackle the question we asked earlier, namely how to figure out whether a particular fly could hold its liquor (or not).

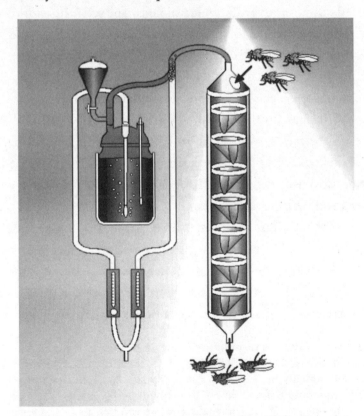

Figure 1.2. The Inebriometer. Courtesy of Dr. Ulrike Heberlein,
Janelia Laboratories. Contraption and flies not to scale.

The first version of the instrument was a glass tube about four feet tall and three inches wide, containing a number of equidistant plastic "steps." The apparatus was set in an upright position (long side up) and a number of flies were put in.

The normal behavior of flies in any enclosure is to fly upward to the top. Sure enough, they did the same in the inebriometer, and this behavior is a key factor that allowed the idea to work. Once the flies were all the way up to the top, clinging to the walls of the glass tube, an amount of vaporized alcohol was introduced into their midst. As you can imagine, drunken flies had a harder time holding on to the glass surface at the top of the tube and would oftentimes slip and fall; understandably, there was a direct correlation between how drunk the fly was and how soon they lost their footing. The plastic steps would catch the falling flies, giving them a chance to rest and regain their stability. The idea was that the lower the step at which a fly regained its footing, the drunker the fly. By running the experiment through several iterations, researchers were able to separate the flies into different populations based on their sensitivity to alcohol.*

As intended, the inebriometer proved helpful in selecting populations of flies that differed in their sensitivity to alcohol, and evolutionary biology types like the inebriometer's inventors were able to apply this information to their research. However, this contraption also allowed for the design of beautiful pharmacological experiments, and pretty soon other scientists were inspired to further modify the device to measure various behaviorally-based drug responses.[43]

Of course, as we know, our curiosity about how alcohol affects animals is not a modern notion, and there are examples dating from the 1800s of efforts to quantify animal "drunkenness." Dr. William

---

* In fact, the inebriometer is useful to assess the anesthetic properties of any type of volatile molecule, not just ethanol (and actually, ethanol is a weak anesthetic; it only works as such in relatively high amounts). For an example of the use of the inebriometer to explore other anesthetics in *Drosophila*, please see Dawson and collaborators (2013).

Lauder Lindsay, a physician/botanist we will encounter in chapter four when we talk about his paper "Mind in Plants," also did some pioneering work on alcohol psychopharmacology. His 1879 book, *Mind in the Lower Animals in Health and Disease*, is a five-hundred-plus-page exploration of the differences (though mainly illustrating the similarities) between animals and people, including those in the realm of behavior.[44] Among many other curiosities, Lindsay documented a good number of cases of intoxication in animals (to be fair, in quite a few of those cases, he was the instigating cause), and generated what I believe is the first-ever example of a scientific *inebriation scale*. The scale uses observed behavior as a measure of how intoxicated an animal is, beginning at 1 ("simple excitement"), passing through 5 ("eccentricities of motion"), and finalizing in 10 ("death"). Lindsay's scale is obviously a less rigorous drunkenness-measuring methodology than that used in the inebriometer (while "death" is likely relatively easy to pinpoint, "eccentricities of motion" seems like a tricky variable), but it underscores science's long-standing interest in quantifying how psychoactive substances affect our animal cousins. Quite another question, of course, is why they seek them out in the first place.

## DRUGS AND DRIVES

In 1989, well over a century after Darwin corresponded with his colleagues about the affinity of various koalas and monkeys for tobacco and alcohol, Dr. Ronald K. Siegel articulated a rather provocative hypothesis about why organisms seek out psychoactive drugs. The late Dr. Siegel was a bit of a maverick (we'll hear much more about him in later chapters), and his bold proposal was that drug seeking is a universal, fundamental drive in all animals, not just in humans. Further, he believed that this drive was equal in magnitude to thirst, hunger, and reproduction.[45] Dr. Siegel's hypothesis is disputed with a certain degree of, let's say, "intensity" in the pharmacology field in general and the psychopharmacology field in particular, and it is not hard to imagine why. The most obvious objection to this idea is that while drug seeking can indeed be a strong drive in the animal kingdom, we have no evidence of these behaviors being universal, much less as fundamental as seeking nourishment; it is possible to develop or even get rid of an addiction, but the need for—and drive to seek out—water or food is ever present (the mating drive, too, is present in all sexually reproducing species). An additional consideration that casts doubt on Dr. Siegel's hypothesis is the degree of variability within a particular population in terms of how drugs affect the behavior of an organism. That is, as we saw with the inebriometer, not all members of a group of organisms of the same species will react the same way to the same drug. These various reactions induced by the same drug in different individuals are significantly influenced by any number of genetic factors, and they go far beyond the level of individual variance we see in hunger or mating drives. After all, if an animal possesses mutations in the genes that control hunger or the reproductive drive, said individual may have little hope of surviving, let alone reproducing. In contrast, the craving for psychoactive substances can often be overcome with minimal effects on survival or reproduction. This fact alone strongly argues against the idea of a "universal drive."

The aforementioned "universal drive" is but one of *two* provocative hypotheses put forth by Dr. Siegel on the topic of why organisms seek psychoactive drugs. His other idea is less controversial (and most likely correct), namely that animals acted as literal role models for humans in our discovery of nature's chemical bounty—that is, that noticing the effects of certain plants on animals led us to experiment with them ourselves. Dr. Siegel called this notion "my theory" in his 2005 book, *Intoxication: The Universal Drive for Mind-Altering Substances*. However, at the time of this writing, I have been unable to find any academic paper formally supporting the notion that he actually formulated and published this theory before anyone else; perhaps I have not dug deeply enough. At any rate, throughout the book, our conversation will explore these ideas further (and those of a few other "mavericks" along the way).

Whether or not one agrees in part or wholly with Siegel's hypotheses, it certainly is true that, throughout history, humans slowly but steadily built serious relationships with certain plants and fungi, relationships that spanned millennia and perhaps millions of years and are still going strong. In all likelihood, these plants and fungi first caught our attention as sources of food, and it was only later that we learned of their other properties, including their psychoactive effects. We found some of these properties harmful, some of them useful (as in medicinal), and some others merely desirable (as in recreational). Sometimes we discovered that the same compound could be all three, depending on how we took it and how much of it we took. Over time, as we unraveled more of nature's secrets through science, we learned that these characteristics were due to specific chemicals or groups of chemicals these organisms produced. Once we learned some chemistry, we were able to "play" with these substances; we haven't stopped playing since.

There are many available resources listing the chemical and pharmacological characteristics as well as the physiological effects of the most commonly known drugs, so I will not go over the generalities. However, an important detail about the drugs we will talk about in this book is that they can generally be divided into two categories: alcohol (ethyl alcohol, the sole member in this category) and a variety of nitrogen-containing tox*ines*.* Many—if not most—of these *toxines* are alkaloids (more on this later). The alkaloids cocaine, nicotine, morphine, and caffeine, among other "-*ines*," have played important

---

* Though I am a prolific typo generator, this is not a typo; I am aware that the correct English spelling of this term is "toxins." My whimsical choice of moniker reflects two facts: most of these compounds act as natural insecticides/pesticides (and are therefore toxins) and virtually all of these compounds possess a nitrogen atom in a special configuration making them *amines*. Hence, *toxines*.

roles in human societies, and the stories behind the discoveries of these compounds are closely intermingled with our history. And human history has long been—and still is—largely the history of trading and economics. Once we invented money and began to use it to trade for goods and services, it was only natural to apply this social tool to obtaining psychoactive substances in significant, marketable amounts, finding ways to profit from pleasure via booze and the toxines.* Even now, ethanol and many types of products containing psychoactive compounds power up a significant fraction of the world's economies, legally or otherwise (think alcoholic beverages, chocolate, tobacco products, abused and illegal drugs, and tea and coffee).

Over time, many factors coalesced to deem certain substances "licit" or "illicit," and surprisingly often these factors had nothing to do with whether a specific compound was especially dangerous or even psychoactive. In more than one case, societies labeled substances as "good" or "evil" based solely on economic factors. Though I try hard not to be judgmental, I cannot help but feel amused by the fluidity of what we consider "good" or "bad" (which often translates to "legal" or "illegal") depending on the historical period, especially in the case of psychoactive substances. As an example, the enthusiastic embrace of tobacco by most of Europe (and eventually the rest of the world), a trend that began in the 1500s, could just as easily have been enjoyed by coca plants. Some authors argue (I think correctly) that it was simply a matter of which plant happened to arrive in Europe first. And while chances are good that most of my readers start their day with a cup of coffee, it might surprise you to know that at various times during history, caffeinated drinks were deemed harmful, illegal, and even evil, a state of affairs that—thankfully—did not persist.[46]

---

* Wouldn't "Booze and the Toxines" be a great name for a band?

## A CUP OF CORRUPTION

In 1777, Frederick the Great, king of Prussia, offered the following royal pronouncement: "It is disgusting to notice the increase in the quantity of coffee used by my subjects, and the amount of money that goes out of the country as a consequence. Everybody is using coffee; this must be prevented. His Majesty was brought up on beer, and so were both his ancestors and officers. Many battles have been fought and won by soldiers nourished on beer, and the King does not believe that coffee-drinking soldiers can be relied upon to endure hardships in case of another war."[47]

Alas, while His Majesty couched his opinions in concern for the fitness of Prussia's fighting force, the august ruler had a political agenda when giving this impassioned speech from atop his proverbial or even literal high horse (history is silent on the details of this fact). The growing popularity of coffee as a beverage (a beverage, unlike beer, not produced by Prussia) was threatening Prussia's economy at the time. Thus Freddie patriotically denounced coffee with a passion—while also putting in a good word for beer. Coffee drinking was a disgusting habit; beer, though, was more than okay—for children, even, beer was good; the king had spoken!

It is difficult to imagine a world in which the ubiquity of tobacco was replaced with the ubiquity of cocaine—and in which we begin our modern mornings with a foaming mug of beer. However, the cultural popularity of various substances is, well, cultural, at least to an extent. But throughout time (and hardly surprisingly), virtually all of the most popular substances have been those that have (pleasant) psychoactive effects, at least when taken at reasonable concentrations.* It is clear that the motivation of our ancestors for consuming medicinal plants went beyond these plants' ability to alleviate maladies. Like us, our animal cousins consume substances that display evident beneficial health effects, but they also consume other substances that do not seem to be good for anything except inducing behaviors suspiciously similar to intoxication. This is, of course, precisely the reason so many of us actively seek out these substances, a series of behaviors that in turn contributes to the general phenomenon of addiction.

Alongside many multicellular living organisms, we have a natural, built-in tendency toward addictive behaviors simply because of how our nervous system works. Our neural architecture is directly responsible for drug-induced reward phenomena (more simply put, pleasant feelings in response to consuming a drug), and therefore for our tendency to engage in these behaviors. Plant-produced chemicals mimic neurotransmitter action, which causes a variety of responses, including the aforementioned pleasurable sensations.†

While these effects and behaviors, as we have and will see, are ancient ones, the formalization of the concept of addiction is less than a century old. Before we understood the neurobiological basis of addiction, society at large harshly judged people who were addicted to anything, labeling them as lacking morals, self-control, or both. Today, though stigma persists, addiction is widely recognized as a disease with an underlying biological basis. This is good news, not least because it raises the hope of finding medical solutions to the widespread phenomenon of substance abuse.[48]

The specific type of addiction we will refer to in this book is the chemical variety, namely the active and persistent seeking of

---

* Of course, in this context, the word "reasonable" is very subjective, as we will see in the "Philosophy of Pharmacology" section in the next chapter.

† In virtually every case, these neurotransmitter-mimicking compounds—like the neurotransmitters they imitate—are alkaloids. We will delve into the topic of alkaloids in a later chapter.

substances that induce psychoactive effects. A formal definition of addiction from the National Institute on Drug Abuse is a "chronic, relapsing disorder characterized by compulsive drug seeking and use despite adverse consequences."[49]

It hardly needs saying that "addiction" is a frequently misused term, and we use it colloquially to refer to all sorts of behaviors that are not directly related to chemical abuse. For example, we talk of people being addicted to gambling, sex, working out, toxic relationships, smartphones, and even gossip—if you can imagine an interest, you will surely find people "addicted" to its abuse. Although it is true that brain chemistry surely plays a part in the abuse of any pleasurable activity, we will sidestep the debate around what constitutes addiction more generally and keep our focus on the direct pursuit of psychoactive drugs.

One thing is certain: all multicellular organisms displaying anything remotely analogous to what we consider a nervous system (that is, having a brain or even a brain-like structure) are in principle capable of displaying addictive behaviors (albeit with the aforementioned differences accounted for by genetic variation) to something. Pretty much every substance that we mention in this book has the potential of becoming the object of desire of a creature—human or otherwise—suffering from addiction, a phenomenon that expresses itself in multiple ways and, as has been well documented, displays many layers of complexity.[50] And it does not matter whether we are talking about vertebrates or invertebrates, wombats or worms; research tells us that addiction-like behaviors in virtually every type of organism ever examined are heavily influenced by the same neurotransmitter that mostly controls addictive behaviors in humans: dopamine.[51] As you may imagine, these commonalities are the basis for our exploration of the effect of drugs on our nonhuman cousins.

---

As this chapter has made abundantly clear, both people and animals have been intentionally consuming psychoactive substances for a long time, beginning with alcohol. You may recall that there is strong circumstantial evidence of purposeful fermentation (that is, induced by humans) in order to produce alcohol as early as about one hundred thousand years ago. However, in all likelihood, humans first noticed the phenomenon of alcoholic fermentation in a serendipitous way—that is, entirely by accident. A possible scenario tells the story of a hypothetical ancestor who was hungry enough to eat some wet, smelly, fermented grain:

*The bitter taste of the fermented grain was not particularly pleasurable, but hunger eclipses virtually every other possible drive and preference, and so Glaarg ate it anyway. And after a period of time, when his hunger had subtly subsided, our hero became aware of certain unexpected sensations. There was a warm feeling—both literally, in the chest area, and figuratively, toward the world at large—as well as a certain degree of not-unpleasant sleepiness. Later, contemplatively banging one rock against another, Glaarg recalled these sensations, and thought about how he might experience them again.\**

The above scenario may be imaginary, but it is not far-fetched, nor is it far-fetched to imagine that similar scenarios could apply to the discovery or invention of a variety of fermented meals in various cultures. Of course, in this context, one of the essential requirements for fermentation is the presence of some common fungi we know as yeast and, strangely, it is possible this contribution brings us back around to our friend the fly. Some researchers have hypothesized (to be fair, this is little more than speculation) that the initial accidental "inoculation" of yeast into a grain mixture intended for nourishment might have come from the addition of ripe, yeast-rich fruit that kick-started the fermentation process. And another hypothesis states

---

\* This little scene is entirely the product of my own imagination, but the *italics* give the story a dramatic touch, maybe even some *gravitas*, don't you think?

that said yeast could have found its way into the grain from none other than the bodies of fruit flies that were attracted to the fruit's smell, landed on it, and "infected" the fruit with alcohol-producing yeast. Talk about serendipity!

Alas and again, as with so many biological mysteries, this one is likely to remain unsolved; we will probably never know whether the humble fruit fly really put the "buzz" in "buzzed." But I like to think it may have.

## CHAPTER 2

# SOME OF THE USUAL SUSPECTS, AND A FEW OF THEIR STORIES

*There is hardly a more difficult chapter in the whole of pharmacology than an exhaustive and thoroughly exact analysis of the effects of drugs.*
—DR. LOUIS LEWIN, *PHANTASTICA*

Whether by their own devices or on the heels of some adventurous animal (more on these later), our ancestors came into contact with myriad species of plants and fungi with psychoactive effects. As with ethanol, some of those effects were deemed desirable, even pleasurable, and as a result, the substances that produced them became sought after. Throughout history, many cultures recorded the properties of these plant/fungi–derived substances; however, the first treatise attempting a comprehensive catalog of the various psychoactive effects of the consumption of these organisms did not appear until 1924: the book *Phantastica* by Dr. Louis Lewin, a German physician.[*][1] *Phantastica* was ahead of its time in many ways. For example, Lewin acknowledged that while not every substance that affected thought processes was hallucinogenic, every such substance is psychoactive in some way. Oftentimes, we equate the terms "psychoactive" and "hallucinogenic," yet even good old aspirin can be considered psychoactive, if only when we take into account the change in mood we experience when a headache gets better. What's more, Lewin implicitly took what we think of today as a "pharmacogenetic approach," meaning that he discussed the different effects the same substance may induce in different people (for example, the higher susceptibility of some people to becoming inebriated) as well as the expression of phenomena like tolerance and habituation, among others.

Lewin classified psychoactive substances into four main categories: *Euphorica* (sedatives, like poppy plants); *Phantastica* (hallucinogens, like magic mushrooms); *Inebrianta* (alcohol; enough said); and *Excitantia* (tobacco, coffee, cocaine, etc.). He lamented the fact that, when he wrote the book, the science of chemistry was not advanced enough even to analyze, let alone synthesize, the chemical principles responsible for the effects of these substances on the human mind. In his words: "No chemical research has been able to produce synthetically anything in the slightest degree resembling the materials which peoples of all parts of the world have found suitable for their euphoric cravings."

As we know, that is no longer the case.

In this chapter, we will introduce some of the drugs other than alcohol that make up our cast of (nonanimal) characters, and explore

---

* To be clear, Lewin's is the first published work *we know of* that systematically explored the world of psychoactive plants and fungi. In all likelihood, there were others that are lost forever due to the criminal destruction of ancient libraries. Think about the Library of Alexandria or the burning of most Mayan codices by the Spanish clergy, among many other examples. Thus, whenever I say "the first" (unless I say otherwise), I will generally mean "the first *known*."

curious facts and further intriguing thoughts about the history shared by animals (us included) and these substances. But before we begin, the time is beyond ripe (you might say it is close to fermentation!) for a bit of grounding on the subject of drugs more generally.

## THE PHILOSOPHY OF PHARMACOLOGY*

It is unusual to see these two academic terms in the same sentence, as pharmacology and philosophy do not seem to be related to one another. However, each scientific discipline in fact has its own set of guiding philosophical principles, and pharmacology is no exception.

The first order of business is to define what exactly we mean when talking about a "drug." As with many important concepts in life, this term is subjective, meaning that its precise meaning will depend on whom you ask. Nonetheless, we can say for sure that at the fundamental level, a drug is a chemical, and a common definition of a drug is "a chemical used to treat a disease." This being said, there are many compounds that are not medicines, but which we still classify as drugs. Take the example of nicotine, which is probably the most addictive substance known. This drug is extensively abused by humans, yet I know of no *bona fide*, medically approved use for nicotine so far.[2] Perhaps a better way to define "drug" would be to differentiate it from food, as few of us drink martinis for their nutritional value; however, just as defining drugs as "medicine" isn't quite right, defining them as "not food" has its own problems, which we'll delve into more later. For our working definition, then, perhaps the best we can do is something like "a non-nutritional substance or chemical that affects the function

---

* Please note that this is not an "official" integration of these disciplines. I have to confess that for a nanosecond I considered titling this section "Pagán's Rules of Thumb," but I thought that was a bit much, perhaps. At any rate, the point is that, while this formulation is my own, these concepts are not merely my invention; they are widely used in biomedicine.

of your body." At any rate, we call the branch of science that studies drugs—whether medicinal or recreational—*pharmacology*. Like most modern scientific subjects, pharmacology is interdisciplinary in nature; it has to be, because in its modern inception, this science deals with the effects of drugs on humans and animals, from biochemistry to behavior.

## FROM THE PLANT TO THE DRUGSTORE, VIA THE CHEMIST'S FLASK

When a drug is just beginning to be studied, it is usually classified in one of three categories: "naturally occurring compound" (this is fairly self-explanatory), "semisynthetic" (chemically modified versions of naturally occurring products), and "synthetic" (meaning that based on the best available evidence such a compound has never been observed in nature). One of the best-known examples of semisynthetic drugs is aspirin, which has a distinguished history as one of the very first widely commercially available drugs. The chemical name of aspirin is *acetylsalicylic acid*, a derivative of the natural product *salicylic acid*, commonly found in willow trees as well as in other plants and algae. Aspirin is just an example; there are many others. In fact, most nature-inspired drugs currently in clinical use are semisynthetic.

Almost without exception, pharmacologists (including yours truly) rely on four core principles that guide our work and help organize our thoughts and ideas. These apply whether the context is one of experimentation or of established drugs used in clinical practice, and keeping them in mind will help us understand what is going on as we continue to explore the world of animals on drugs, and drugs more generally.

## 1. Cause and Effect = Dose and Response

Depending on the context, we can call this principle the *dose-response concept* or the *concentration-response concept*.[3] This means exactly what it sounds like: for any given drug, the induced effect is proportional to the amount given (but please see principle #2!). In other words, the higher the drug dose, the larger the effect and vice versa. Of course, this is true whether we are talking about a drug's desired effect or other, less desired consequences of its consumption. Too much of a good thing oftentimes becomes a bad thing, which brings us to our second principle.

## 2. There Is a Fine Line Between a Drug and a Poison

About five hundred years ago, the Swiss scientist Paracelsus gave us these wise words: "*Sola dosis facit venenum.*" For those of you whose Latin is a bit rusty, this roughly translates to "the dose makes the poison." You've probably heard this phrase before; it is oft-quoted precisely because it is so important. This principle tells us that *no substance is 100 percent safe*, meaning that if you take any substance—and I mean *any* substance, even water—in excessive amounts, said substance is likely to become harmful. We usually refer to this harm as *toxicity*.[4] The concept of toxicity was the origin of pharmacology's evil twin: toxicology. This discipline essentially studies the pharmacology of a drug's harmful effects or, as I call it in my silliest moments, *when pharmacology attacks.*[*]

It is critical to understand that in the medical sciences the decision to administer a medicine is largely a matter of determining the cost/benefit ratio. In other words, doctors give a medicine when the small probability of harm is more than offset by the likely benefits (or by the certainty of harm if nothing is done to prevent or treat a given

---

[*] Of course, toxicity also deals with substances that we do not generally associate with consumption for *either* medicinal or recreational effects, like certain heavy metals or cyanide, for example.

disease).* Of course, while this cost/benefit ratio applies on a population level to drugs we decide are safe enough (at a particular dose) to
be used as medicines, it is applied by doctors on an individual level,
too: a particular drug might not work equally well for everyone, while
on the other hand not all substances are equally toxic to everyone (not
everyone is allergic to penicillin, for instance). After accounting for
factors such as age, general health state, and so on, the varying effects
that the same compound might induce in different individuals are
largely a matter of genetics, which brings us to our third principle.

## 3. We Are All Different

Populations are almost never made up of clones; in nature, a clonal
population is generally not a very good idea because variation is one
of the main prerequisites for evolutionary change and adaptation.[5] As
for how this relates to drug effects, the main idea is this: in any normal population of a particular species, all individuals will have a basic
genome that defines their particular species. For example, all seven
billion or so of us humans on this planet right now share the same
basic set of genetic material (what we call the human genome). However, with the exception of identical twins, triplets, and so on, there
exist many individual differences going "beyond" that basic genome.
This is self-evident; just think about all the physical differences in
any group of people—aside from the aforementioned exceptions,
everybody looks a bit different, while still having the general arrange-
ment of anatomical features that are the hallmark of what we know
as "human." Applying this reasoning to our genes, there is a series
of genes that define us as being human; there is little wiggle room
in these (for example, no normal humans—except for moms—have

---

* One example of this idea is the use of certain steroids to treat asthma.
Several of these steroids' side effects include elevated blood sugar, which can
be a concern in many people. However, during an asthma attack, the most
pressing concern is restoring the ability to breathe, not a transient elevation of
blood sugar.

eyes on the back of their heads). Other genes, like those that control the color and texture of hair (red/brown/black/blond hair or—in my case—no hair at all), can "come" in several variations. The genes that may affect our responses to drugs fall into this latter category. Let's explore this concept a little further.

Take three people who, sadly, are each suffering from a headache (and to make things simpler, let's imagine their headaches are all the same). Perhaps the first person only needs one regular aspirin to get rid of the pain, while the second one needs two of those, and aspirin does not even work for person number three, but another drug does the trick. This possibility of variation applies to virtually every medicine at our disposal, and it's the same with poisonous substances. Prolonging our hypothetical headache (it'll be over soon, pharmacologist's promise!), suppose that we have a large group of sufferers and a supply of standard 325 mg aspirin tablets. Many of the individuals in this population will be able to rid themselves of their headache with only one tablet. But a certain percentage of the population will need much less than 325 mg to feel better, while another fraction of the population will require more. There might even be a fraction that is either completely insensitive to aspirin or even poisoned by the smallest amount of it.

Animal populations are no different. Please keep this idea in the back of your mind as we continue our conversation! For some reason, we humans have a tendency to assume that all animals of a certain type are pretty much the same—we are less likely to do this with familiar creatures we are used to thinking of as individuals, like dogs, but when we think about, say, koalas, we often imagine them as a bloc. However, just as we discussed for human populations above, animal populations also display a high degree of genetic variability. Thus, if we talk about a certain animal species' sensitivity to a particular drug, it will almost never mean that 100 percent of the individuals of such a species will be identically affected. For example, later on we will talk about the active chemical in catnip and its effect on, well, cats.

As we will see, only about 70 percent of individual cats seem to react to catnip. This fact does not only apply to house cats, mind you, but bigger cats as well (think tigers and lions). We may not have this kind of detailed information for every animal species that we encounter in this book, but if I have found it and it is relevant to our conversation, I will mention it. The bottom line is that, for all creatures great and small, some individual variability in response to a drug is not surprising and even expected.

## 4. Everything Begins with Binding

Until now, I have not said too much about *how* drugs are able to affect biological systems. To illustrate the mechanism behind this process, let's think about how we hear music. We cannot capture the music carried by electromagnetic waves traveling through space at the speed of light without an appropriate device such as a radio.* Similarly, for an organism to react to a particular drug, it needs a means of detecting the presence of the chemical. No chemical is capable of inducing a specific biological effect without physically interacting with a specific molecular "antenna," usually located at the surface of certain cells. We call these antennae *receptors*, and the interaction between a drug and its receptor is called *binding*.†

In essence, a receptor's job is to react to the presence of a chemical when said chemical comes in contact with it (here we'd say that the chemical *binds* to the receptor). As a result of this reaction, something happens (muscle contraction, hormonal signaling, and so forth). In

---

* If you hear music in your head—that you are not thinking about—without the help of a radio or any other kind of electronic device, please see a licensed health professional.

† In other words, *everything begins with binding*. This is my free translation of the famous (among pharmacologists) maxim of the father of receptor theory, Paul Ehrlich: "*Corpora non agunt nixi fixata*," which pretty much means that "chemical entities cannot interact with each other unless they are in physical contact." For example, a local anesthetic must bind to certain parts of the nerve cell in order to prevent pain.

general, receptors are proteins that control biological responses; their specific nature depends on what they do in the organism. The formal pharmacological definition of a receptor implies a protein to which a chemical binds, making something happen. However, there are many variations on the theme—for example, transporters (which do exactly what it sounds like they do), enzymes (essentially the molecular entities that perform the actual work in a cell, like controlling chemical reactions), as well as quite a few other classes of receptor-like molecules. So when your dentist gives you a local anesthetic, the anesthetic molecules bind to a population of specific proteins on the surface of your nerves, inactivating them (hence, no pain). In this case, the protein receives the substance, but instead of activating the protein and making it do something, this binding turns it off.

The point is that any drug that enters the body interacts with a target protein (or more likely proteins) in exactly the same way than a natural neurotransmitter or hormone will, by binding. In some cases, the drug will mimic the action of a natural substance (in this case we call it an "agonist"); in some other cases, it will prevent such action (we call these "antagonists"); still other times, we might be talking about "modulators."

What are modulators? I'm so glad you asked! Imagine a light switch. Ordinarily, we might turn it "on" (as when an agonist activates). We can also turn it off (as when an antagonist inactivates). Or we might have a dimmer, which we can manipulate to make the light bulb brighter or fainter. That's a modulator.

(Can you tell that I am an actual pharmacologist? I could talk about these things until the proverbial cows come home, but let's get back to the point, shall we?)

From now on, unless I tell you otherwise, whenever we talk about a particular substance you may assume that said substance interacts with a specific receptor—more often than not, with a family of such receptors. There are receptors for virtually every substance we will talk about, including hallucinogens, nicotine, morphine, neurotransmitters, hormones, and so on. And in the special case of a drug that interacts with a transporter or an enzyme, I will note it and provide any additional information relevant to our conversation.

---

Now that we are all on the same page, pharmacologically speaking, let's embark on a brief tour of some of the familiar plant-based substances consumed by creatures—our own mini-*Phantastica*.

## POPPIES AND THE GIFT OF JOY

Many of the drugs we'll discuss have a bad reputation, and for good reason. Aside from the obvious issue of addictive potential, virtually every kind of mind-altering substance has the potential to induce frightening hallucinations, paranoia, and other effects usually associated with the expression "bad trip."

Yet—while they are definitely among the worst offenders in terms of addiction—there is one particular group of psychoactive compounds that (at a "reasonable" dose, standard caveats applying) almost always induces "trips" that are pleasant. Sometimes, very pleasant indeed—in fact, these compounds are known the world over for their ability not only to remove pain but to suffuse the mind with pleasure. They come to us courtesy of poppy plants (*Papaver somniferum*), which are of course renowned for producing the opiates (that is, chemicals that come from opium) morphine, codeine, and heroin, among others (Figure 2.1).*

---

* Opiates are naturally occurring products; opioids are semisynthetic derivatives, like heroin.

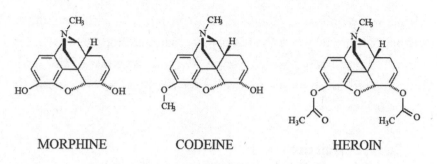

MORPHINE                CODEINE                HEROIN

*Figure 2.1. Morphine, codeine, and heroin. Drawn by the author.*

The plant's scientific name gives us a hint as to its effects, as *somnium* is the Latin word for "dream." Moreover, poppy plants and their close relatives (of which there are about a hundred or so) have been humanity's companions from time immemorial,[6] and the names given to them by various cultures in prescientific times were no more subtle. The Sumerians called the poppy *Gil* (happiness or joy) or *Gil-Hul* (the joy plant), and early Arab culture called it *Abou-el-noun* (father of sleep). The ancient Greeks associated it with deities such as *Hypnos* (god of sleep) and *Morpheus* (his son, responsible for dreams).

If ever there were a substance that illustrates the complexity of assigning moral values to chemical compounds, surely this is it. If you've ever had surgery, or even a kidney stone, chances are you appreciate the usefulness of the poppy plant. The development of the host of lifesaving surgical procedures that extend and improve our lives was made possible by morphine—as was the ability to ease the pain of the dying. At the same time, it is no accident that this plant of pleasure is associated with some of the worst ravages of addiction. Whole wars (most notably, ahem, the Opium Wars) have been fought over it, and like most drugs, it has been subject to the whims of societal opinion. Bayer (of aspirin fame) marketed heroin as a cough remedy as late as the 1890s—for children, no less! Interestingly, in the 1600s, the Chinese emperor Tsung Chen made the use of tobacco illegal, while opium remained above reproach.* As with alcohol, the history of humanity and the history of opiates are indisputably inter-twined.[7] For our purposes, however, we can leave most of this aside, touching only on how we discovered the chemical secrets behind the poppy plant's allure.

## The Sleepy Pharmacist

Extracting psychoactive material from the poppy plant is ridiculously easy. The official definition of opium is the "air-dried milky exudate obtained by incising the unripe capsules of *Papaver somniferum*,"[8] but the milky substance containing the relevant compounds is present in virtually every part of the plant. (As you might imagine, this makes it convenient for animals—for instance, some wallabies we'll meet in a later chapter.) Figuring out the chemical entities responsible for the plant's effects proved a tad more elusive, but it was eventually accom-plished in the early 1800s—with the help of animals.

After more than a few attempts at isolating the actual chemical in *P. somniferun* responsible for its narcotizing effects, the German apoth-ecary (an older name for a pharmacist) and chemist Friedrich Wil-helm Adam Sertürner obtained a whitish powdery substance, which suggested a more or less pure compound, and he happily proceeded to call his substance *morphium*, after the aforementioned Greek god. He then (wisely) decided to perform some animal testing, by lacing cheese with the crystals and giving it to mice.† Very inconsiderately, the mice died (likely due to overdose). But Sertürner, undiscour-aged, simply moved up the food chain, so to speak: he coated some bones with his crystals and gave the bones to a few dogs. One of the dogs died as well, but the rest "seemed happy" and were apparently

---

* I wonder if the good emperor's rationale for outlawing tobacco in favor of opium paralleled the outlawing of coffee in favor of beer in the 1700s by our friend Frederick the Great from chapter one . . .

† Stereotype much? Mice actually eat much more than cheese!

unharmed.* Sertürner deemed these results "good enough" and proceeded to expand the experiment to human subjects by ingesting a quantity of the morphium crystals himself.

He woke up ten hours later.

Sertürner was a lucky man; historians estimate he consumed twice the maximum amount of morphine nowadays considered safe. This was not the first time—nor would it be the last—that an adventurous scientist experimented with a compound of their own making by testing said compound on themselves (early chemists—whom you'd really expect to know better—were famous for *tasting* their products). "Auto-experimentation" is an arguably distinguished tradition that has led to the development of many drugs and even earned some of these scientists Nobel Prizes. On the other hand, many of these scientists also suffered harmful effects due to intake of their newly discovered compounds—up to and including, in some cases, death. Depending on the substance, of course, another risk was addiction; we'll see one of the best-known examples of this as we explore our next drug.

## COCAINE: IT STARTS WITH A TOOTHACHE

We've said that our relationship with virtually every type of psychoactive substance began when we discovered curious effects upon chewing, eating, or smoking the plant (or fungus) that produced it, and our shared history with cocaine is a perfect example of this. In South America, likely millennia ago, a person chewed on the leaves of a coca plant† for the first time, probably as a snack. Soon, the adven-

---

* He did not specify what criteria he used to determine the happiness of the dogs.

† The name that the South American natives of the area gave to the plant was *khoka*, which to the Spanish settlers of the area, sounded like *coca*, and the Spanish word stuck.

turous snacker felt a strange sensation on their lips: a not unpleasant tingling and numbing that faded over time. Of course, chewing on coca leaves would, depending upon the dosage, have other effects as well, but let's stay with this one for a moment.

The active compound of coca leaves is, of course, cocaine (Figure 2.2), and while cocaine was not the first widely used psychoactive agent, it is arguably one of the most infamous.[9] For our purposes, what makes cocaine most interesting is its chemical properties, which are responsible for its status as the first *bona fide* local anesthetic.

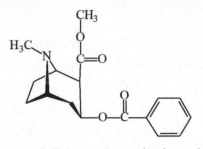

Figure 2.2. Cocaine. Drawn by the author.

While chewing on coca leaves may have been an ancient practice, the earliest written account of their anesthetic effects was published in 1653 by Bernabé Cobo, who was (of all things) a Spanish Jesuit priest living in Peru.[10] I've narrated his tale in more detail elsewhere,[11] but it essentially goes like this: One fine day, Father Cobo got a toothache, and so naturally he promptly paid a visit to his barber (barbers were also the dentists of their time—a bit alarming, I know) to have the tooth taken out. The barber said he would rather leave the tooth in as it was not too badly damaged, and sent the miserable Father Cobo back to the monastery to suffer. A fellow man of the cloth knew that the natives of the area sometimes chewed coca leaves to alleviate gum and toothache pain, and suggested poor Father Cobo try doing the same.

I like to imagine Father Cobo—in pain and at this point likely rather annoyed—raising a skeptical eyebrow at his compadre's "leaf chewing" idea, and perhaps even muttering a rather unholy expletive in Castilian. But, as the saying goes, desperate times call for desperate measures, and after all, what did he have to lose? Father Cobo went ahead and chewed on some coca leaves. To his unexpected yet absolute delight, the toothache got better, and the now much happier priest wrote about his (mis)adventure for the benefit of future generations.

Alas, the use of the main product of the coca plant as an anesthetic agent would not become widespread until a couple of centuries later.[12] History records that cocaine was first isolated in 1855 directly from coca leaves by the German chemist Friedich Gaedcke. A few years later (around 1859 or 1860), a German chemistry student, Albert Niemann, was the first to characterize cocaine in chemical terms (meaning elucidate its precise molecular structure), as part of his doctoral dissertation at the University of Göttingen. There is no indication that either Gaedcke or Niemann had any inkling of the possible usefulness of this compound; alas, Niemann died a year after obtaining his PhD, at the young age of twenty-six.

It was not until 1884 that the medical sciences began to properly notice cocaine, and even then, its eventual medical usefulness was realized only as a by-product of the pleasurable sensations it induced in humans. Furthermore, in the time-honored tradition of self-experimentation, these pleasurable sensations were documented—in writing—for the first time by several scientists who tested these substances on themselves. The famous neurologist turned psychoanalyst Sigmund Freud was one of these scientists, and has the dubious claim to fame of being the first public person addicted to cocaine.

The relevant insight into cocaine's anesthetic potential did not come from Freud, however, but from his associate and friend (and later still, frenemy) Dr. Carl Koller, an ophthalmologist. As the story goes, Koller and Freud were using cocaine recreationally, an activity they practiced with a certain frequency. One day Koller noticed that, when tasting cocaine, his lips went numb. He mentioned this effect to Freud, who noticed it, too, but either failed to recognize the significance of Koller's observation or was just too out of it to care. The evidently more alert Koller, however, saw this curious cocaine-induced effect in light of his expertise as an ophthalmologist. As the biologist Louis Pasteur famously said: "Chance favors the prepared mind,"* and Koller was primed to grasp the implications and possible applications of cocaine use as a local anesthetic, as one of the procedures he frequently performed was cataract removal.

Imagine you are a patient being treated for cataracts here in the modern era. Your eye(s) must remain open, so you are fully awake and aware for this procedure, and after your eye is fully anesthetized, you observe in (blurred) detail your physician moving toward your eyeball with a sharp surgical instrument. This is unnerving enough. But if you were having cataract surgery in the 1800s, not only would you *see* your doctor advancing upon you with a scalpel, but once he got there, you would also *feel absolutely everything* he was doing. General anesthetics were only beginning to be used in the 1800s,† and there were no local anesthetic agents available, none at all. A description of anesthetic-free cataract surgery from the perspective of the patient comes to us courtesy of the famous British novelist Thomas Hardy (as reported by Dr. Howard Merkel of the University of Michigan Medical School).[13] Hardy, in a rather uncharacteristically (yet absolutely understandably) "unlyrical" manner, quite accurately stated that "it was a like a red-hot needle in yer eye while he was doing it."

---

* If I had to express my scientific philosophy in a phrase, this phrase would be it. I stated so in my PhD dissertation, as well as in my first book, when describing how my current research program came to be.
† Ether and chloroform were available, but they usually induced nausea and other undesirable effects for eye surgery, for which, as you can imagine, you need to be very still . . .

## ANESTHETICS AND THE
## HISTORY OF EVOLUTION

This historical lack of anesthetic agents has some direct bearing on the history of the theory of evolution, that most famous brainchild of our friend Charles Darwin. As a matter of family tradition, Darwin was "destined" to become a medical doctor, like his father and paternal grandfather before him.[14] Darwin was never terribly interested in a career in medicine, but he attended medical school anyway, chiefly because his father did not give him much choice in the matter. Surgical procedures were developed well before anesthetic agents, at least in Europe and nearby regions, and in those days a typical surgery involved several strong men holding the patient firmly in place while the surgeon worked as fast as possible. It was as gruesome as it sounds, and this practice gave rise to the event that served as the proverbial final nail in the coffin of Darwin's medical career: He witnessed an anesthesia-free surgery on a child. That did it for Darwin. He left medical school, and eventually began to walk the path that led to the theory that links all of biology in a majestic tapestry.[15]

But back to Koller: Having hit upon a possible way of minimizing his patients' pain, he arranged for a public demonstration of his idea, and once the medical world learned about it, the concept of a local anesthetic was happily embraced by physicians and dentists alike. The use of cocaine as a local anesthetic agent became common in the late 1800s, and the fame and recognition that Koller rightfully earned by introducing the clinical use of cocaine to the world rubbed Freud the wrong way—more precisely, he was insanely jealous of Koller. Freud had become an advocate of cocaine for non-anesthetic applications, among them depression and (ironically enough) as a treatment for morphine addiction, and the same year as Koller's demonstration, Freud published some of the results of his months of self-experimentation in the monograph *Über Coca* (which translates as "Super Coca," should you wonder what stance the book takes). He noted the anesthetic effects of cocaine on skin and mucous membranes[16] in what amounts to hardly an aside in the last paragraph, obviously regarding this clinical property as secondary to its psychoactive potential. Freud's pettiness over Koller's celebration went further. He sent a copy of his *Über Coca* to Koller, signed in the following way: "To my dear friend, Coca Koller, from Sigmund Freud."[17] "Coca Koller" was *not* meant as a term of endearment, but to ridicule Koller by painting him as a nonserious scientist. The nickname mortified Koller until the day he died. Indeed, cocaine enjoyed a brief vogue as a general nerve tonic, even making its way into the initial recipe for Coca-Cola,[18] but once the adverse effects became better known, the backlash was swift, and by the end of the century cocaine had fallen out of favor with the medical establishment. The one application that retained its aura of legitimacy was as an anesthetic (it would lead to the development of a drug you may be familiar with, novocaine), and you can imagine how Freud felt about *that*. Over the years, Freud tried unsuccessfully to belittle Koller's role in the discovery, and some four decades later he was still smarting, even blaming his own wife for "distracting" him from his cocaine studies (not a smart move to publicly criticize your wife, it seems to me, but to each his own).[19] It seems to me he might have benefited from a good therapist.

One final character deserves mention in the story of how coca became cocaine. It is likely that Freud first learned about coca from Dr. Paolo Mantegazza, an Italian physician and scientist.[20] Like Freud, Dr. Mantegazza was a neurologist, and a true psychopharmacology pioneer with an interest not just in cocaine but in drugs in general. Between 1858 and '59, a full quarter century before Freud's famous "cocaine papers" and Koller's formal anesthetic applications of the substance, Mantegazza published an account of his experience upon chewing two coca leaves. He noted strong physiological effects including elevated pulse rate as well as colorful hallucinations. In his papers, he was quite clear about the amount he consumed, yet in his words:

> *Some of the images I tried to describe in the first part of my delirium were full of poetry. I sneered at the poor mortals condemned to live in this valley of tears while I, carried on the winds of two leaves of coca, went flying through the spaces of 77,438 worlds, each more splendid than the one before.*[21]

## Keeping Llamas Alert, and the Moth That Defies the Coca Plant

Dr. Mantegazza may have been an outlier, as most members of the early South American cultures that made a practice of chewing on coca leaves do not appear to have done so in order to fly through the splendid space of other worlds, but instead to combat not just pain but also hunger, fatigue, and altitude sickness in this one.[22] Coca's energizing and appetite-suppressing effects are likely behind the ancient practice in certain of these cultures of feeding coca leaves to llamas, an important pack animal in these cultures. Despite the lack of historical records to properly determine just how long coca leaves were fed to llamas, it is likely that people began to do so shortly after experiencing the tonic effects for themselves, and there is archaeological evidence indicating that humans began consuming coca leaves as early as about eight thousand years ago. This practice thus gives us an early illustration of the effects of (kind of) cocaine in certain animals. However, it is interesting to observe that a certain minuscule

---

\* Did he actually count the worlds?

creature can ingest relatively greater amounts of coca leaves and remain entirely unaffected.

*Eloria noyesi* is a nondescript, light-brown, fragile-looking moth with a wingspan barely a couple of centimeters across. The species lives in South America and its common name, the *coca tussock moth*, gives us a hint about its most remarkable characteristic: *E. noyesi* feeds almost exclusively on coca leaves, without any apparent behavioral or toxic effects. In fact, *E. noyesi* is considered an agricultural pest to coca growers, with "outbreaks" of this moth causing the destruction of coca plantations—in one case, the caterpillars destroyed close to fifty thousand acres of coca plants.[*23]

As we will see in an upcoming chapter, some plants—coca among them—have the ability to produce specialized metabolites that act as natural insecticides. Quick preview: plants produce defensive alkaloids, bugs nibble on plants, bugs—sometimes—die. When the bugs do *not* die, their survival as a population is largely due to their ability to coevolve with the plants in their environment to not only become resistant to the pesticide-like substances they produce but oftentimes actively use them to their advantage. Such a strategy is at work in the case of the coca tussock moth: the females lay their eggs on *Erythroxylon* (coca) plants,[24] and upon hatching, the caterpillars—characteristically hungry—feed on the leaves (again, without any deleterious effects).[†] Not only are *E. noyesi's* caterpillars resistant to cocaine's toxicity, but oddly, cocaine retains its chemical identity within the insect after it is consumed. This is remarkable because it implies that the cocaine is not

---

* Starting in the 1980s, there have been suggestions to use *E. noyesi* as an environmentally friendly bioweapon against coca plants. This idea gained some traction when glyphosate, a pesticide regularly used by the Colombian government to spray illegal coca fields, was found to be a possible carcinogen. However, to date, such a plan has not been implemented—I think this is probably for the best, since historically, improperly planned biological interventions to the environment often backfire, but that is a story for another book.
† We will see a similar scenario in the case of the intriguing relationship between monarch butterflies and milkweed plants.

metabolized,[25] as it would be in virtually any other cocaine-ingesting organism, including humans. Furthermore, a certain fraction of the cocaine ingested by the caterpillar is *never* excreted, staying within the organism throughout metamorphosis all the way into adulthood.* I do not know of any studies testing whether the presence of cocaine in the adult gives the insect a survival advantage, but I would not be surprised to hear that it did. To date, the mechanism through which *E. noyesi* is able to resist the cocaine's toxicity and use it for its own purposes is not entirely understood.[26]

## A WELCOME WEED

In my last book, I imagined a scenario in which a group of ancient humans, presumably to keep the cold weather at bay, used dry marijuana leaves as kindling, leading to their inhalation of the smoke—with unexpected consequences. I am certainly not the first person to consider this hypothetical scenario, and it is an entirely plausible story for the discovery of a variety of psychoactive plants. The particular psychoactive compounds *these* hypothetical humans would have inhaled are known as cannabinoids, used for thousands of years as remedies to treat a variety of conditions, as well as for recreational purposes.[27] Another drug that has been notably subject to the shifting whims of human judgment, over the course of history it has both been decried as "demon weed," despoiler of innocent youths, and—as research into its antinausea and pain relief effects progressed—become a common prescription to patients suffering from many diseases, including cancer. In the US over the past few

---

* Which is no mean feat, mind you. Metamorphosis is a fascinating yet poorly understood process through which cells and tissues of the "embryo" (the caterpillar) degrade and reorganize to form the adult animal. If you want to know more about metamorphosis, I suggest a very reader-friendly book: Ryan (2011) *The Mystery of Metamorphosis: A Scientific Detective Story.*

years, many states have legalized marijuana for medical use, and some have decriminalized the drug entirely.

Almost all members of the cannabinoid family are psychoactive compounds present in marijuana plants (*Cannabis sp.*)—the best known of these compounds is THC (tetrahydrocannabinol, Figure 2.3), which is largely responsible for the "high" points of marijuana consumption.

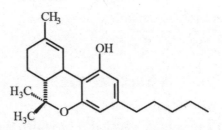

Figure 2.3. THC. Drawn by the author.

There are a few plants besides cannabis that produce cannabinoids, however (best known are some members of the *Echinacea* family), and there are a few cannabinoids that are not generally psychoactive but have other interesting effects. One of the latter is cannabidiol, or CBD, which has recently become omnipresent as a popular remedy for anxiety and pain, in preparations of varying effectiveness. While enthusiasm is somewhat outpacing evidence at the moment, cannabinoids are a varied and fascinating class of compounds and the subject of much research.

Many types of animals, such as amphibians, reptiles, and mammals (including humans), possess receptors for cannabinoids.[28] This is hardly surprising; it is common knowledge to anyone who has visited a college campus that cannabis-derived chemicals affect vertebrates. What is a bit surprising is the presence of cannabinoid receptors in several types of animals we would not expect to have them[*]—among the most unexpected are aquatic animals such as hydra, freshwater planarians, sea urchins, and related organisms (there is no such thing as an aquatic marijuana plant, after all).[29] Of course, it is not uncommon to find biochemical and genetic commonalities between virtually all classes of organisms on Earth, even if they are seemingly useless to some; for evolution, pruning extras is a lower priority than conserving the necessities. What is truly startling is that the best available evidence seems to indicate that insects display a total lack of receptors for cannabinoids,[30] and even in cases where scientists observe cannabinoid-induced effects in insects, these effects do not seem to be receptor-mediated, as in the case of some anesthetics, which simply incorporate themselves within cell membranes without any apparent receptor interaction.[31] Because both "lower organisms" (like planarians and hydra) as well as "higher organisms" (like vertebrates, including humans) possess these types of receptors, it stands to reason that organisms that fall "in between"—like insects—would have them as well. These missing receptors thus represent a true evolutionary mystery. I was similarly unable to find any concrete information about the presence of cannabinoid receptors in arachnids; however, certain spiders produce venom with compounds that act on the cannabinoid system of *vertebrates*. A pharmacological mystery as well![32]

As an aside, there is a persistent (yet unfounded) notion that bees that collect pollen from cannabis plants make cannabinoid-laced honey. There is, alas, no evidence supporting this idea. It is true that many species of bees collect pollen from cannabis plants,[33] but

---

* And yet this is a perfect illustration of how evolution often works. A plausible explanation for the origin of cannabinoid receptors (in fact, for pretty much any receptor) is that these entities were present in ancient organisms for an unrelated "purpose." Over the course of evolution, these receptors integrated with nervous systems, eventually functioning in ways markedly separated from their evolutionary origins. Another example of this is single-celled microorganisms, which do not possess anything remotely similar to a nervous system, yet react to some of the neurotransmitters that control nervous systems.

cannabis plants do not produce nectar and there is no evidence of psychoactive compounds in their pollen. Similarly captivating but unsupported are anecdotal accounts of "grasshoppers that could jump to spectacular heights because of marijuana."[34] In all fairness, grasshoppers can jump quite high without the help of any psychoactive substances.

## THE STRANGEST TRIPS

Many psychoactive compounds are *really* psychoactive; they go way beyond perking you up in the morning. Certain of these—aptly dubbed psychedelics—cause dramatic effects on perception and altered states of consciousness. Even two of the best formal definitions of the term psychedelic, including: "powerful psychoactive substances that alter perception and mood and affect numerous cognitive processes"[35] and "[drugs] with the capacity to reliably induce states of altered perception, thought, and feeling that are not experienced otherwise except in dreams or at times of religious exaltation"[36] do not quite do justice to the interesting nature of the effects these compounds have on our mental states and those of our nonhuman cousins. Of course, the effects are quite difficult to ascertain in the case of nonhuman organisms due to the subjective nature of the psychedelic state, but as we'll see later in the book, this hasn't stopped scientists from trying.

Psychedelics (often in the form of fungi, particularly mushrooms) and humanity are longtime companions, but though we have known each other from before recorded history, our relationship is undeniably . . . complicated. This has resulted in a fair degree of subjectivity surrounding how we name these compounds. Although psychedelic substances are oftentimes referred to as *hallucinogens*, *psychotomimetics*, or *entheogens*, in a strict sense each of these labels describes substances with slightly different subjective properties.

For example, *hallucinogen* and *psychotomimetic* are derogatory terms indicating that such compounds induce pathological states similar to schizophrenia and psychosis, among other states of mind over-whelmingly considered to be negative.* On the other hand, *entheogen* is used to indicate compounds that induce feelings and sensations with mystical and religious meaning, and as such, that these com-pounds are oftentimes used in religious rituals. As you can imagine, these labels are fluid; each can be applied to virtually any type of sub-stance depending on the sensitivity of the person who takes it and the amount consumed. For example, ethanol is not strictly considered a hallucinogen, but it is certainly capable of causing such mental states in sensitive people, especially when "overconsumed."

In the late 1950s, Dr. Humphry Osmond[37] decided to come up with a descriptive yet unbiased term to describe all compounds capable of altering subjective mental states beyond mood or a gen-eral sense of well-being. Essentially, Osmond was thinking about hallucinogenic substances. Some of his choices were, for lack of a better word, peculiar. These included *psychephoric* (mind-moving), *psychehormic* (mind-rousing), *psycheplastic* (mind-molding), *psyche-zymic* (mind-fermenting), *psychelytic* (mind-dissolving), and the memorable—and somewhat dramatic—*psycherhexic* (mind-bursting). To the relief of future generations of pharmacologists, psycholo-gists, and psychiatrists, he finally came up with an unambiguously unbiased and descriptive term that virtually everybody liked: *psyche-delic* (mind-manifesting). The name stuck, and once again, the rest is history. Today, in a sharp departure from the days of associating psychedelics with psychosis, they are being investigated as potential treatments for some psychiatric conditions including depression and anxiety.

---

* These terms literally imply that these drugs imitate or cause psychoses.

Arguably the most famous psychedelic agent, lysergic acid diethylamide (LSD, Figure 2.4), is a man-made substance with a legendary discovery story.

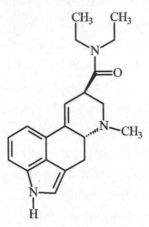

*Figure 2.4. LSD. Drawn by the author.*

In 1938, Dr. Albert Hofmann, a Swiss chemist, was studying certain plants, including *Drimia maritimia* (Mediterranean squill—or, more adorably, the sea onion) and fungi of the *Claviceps* family (ergot), in the hope of isolating various active compounds for use in pharmaceuticals. He isolated LSD from lysergic acid, a natural substance in certain fungi, for potential use as a stimulant, but the compound (formally known as LSD-25) elicited little interest from the "powers that be" at Sandoz, the chemical company where he worked, and so it was forgotten . . . by everyone but Dr. Hofmann, who remained keenly interested in the molecule.[38] At this point, it must be said, no one had any inkling of the psychedelic properties of the compound, yet interestingly, Dr. Hofmann described that he "felt" that LSD was important—call it chemist's intuition, and in 1943, he synthesized a new batch of the chemical. Unbeknownst to him, he got a small amount of the substance on one of his fingers in the process, and a short time later, well, I'll let him tell you:

*In a dreamlike state, with eyes closed (I found the daylight to be unpleasantly glaring), I perceived an uninterrupted stream of fantastic pictures, extraordinary shapes with intense, kaleidoscopic play of colors. After about two hours this condition faded away.*

Chemists being what they are, it probably will not surprise you to hear that Dr. Hofmann responded to this by ingesting a quantity of his new substance intentionally three days later. Forty minutes post-ingestion, he recorded, "Beginning dizziness, feeling of anxiety, visual distortions, symptoms of paralysis, desire to laugh," and decided he had better get home. Most unfortunately, he went by bicycle. It was, by his own account, a very bad trip in every sense.

Once home, Dr. Hofmann sent for milk and a doctor, and proceeded to thoroughly freak out. When the neighbor appeared with the milk, he believed her to be a witch; his furniture seemed to turn on him, and so on. By later in the evening, however, the subject was able once again to enjoy himself:

*It was particularly remarkable how every acoustic perception, such as the sound of a door handle or a passing automobile, became transformed into optical perceptions. Every sound generated a vividly changing image, with its own consistent form and color.*[39]

While later in the book we will hear some stories of nonhuman animals encountering LSD, they generally do so only in experimental contexts, it being a synthetic compound. The psychedelics nonhuman animals are *most* likely to come across—and those with which humans have the longest history as well—are of course natural ones. Dr. Hofmann himself would go on to synthesize psilocybin (Figure 2.5), the active ingredient in "magic mushrooms," and his LSD research, after all, got its start from ergot. So let's look at the latter, a member of psychedelics' "pre-chemical" lineage.

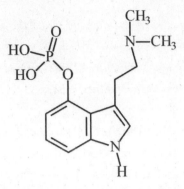

Figure 2.5. Psilocybin. Drawn by the author.

## A Fungus for the Brain

In the introduction, we talked about the discovery of amber samples containing ancient grasses seemingly infected with a *Claviceps*-like fungus. The *Claviceps* genus, known as ergot, counts some sixty species of parasitic fungi that specialize in the infection of grasses (specifically about six hundred species), including some important agricultural crops like rice, corn, wheat, and rye, among others. Oddly, the plants infected rarely, if ever, suffer from any deleterious effects. On the other hand, animals and humans who consume *Claviceps*-contaminated grass suffer a variety of maladies, some of them serious, a fact that argues for a symbiotic relationship between the fungi and the grasses where the plants "allow" fungi to grow in exchange for their chemical "protection."[40] Throughout history, people who ate ergot-infected bread have developed what is known as *ergotism* (also called *St. Anthony's Fire*), caused by a specific species of ergot fungus, *Claviceps purpurea*. Ergotism varies in severity depending on the amount consumed as well as other factors like age and health, but the effects in mammals are widespread. In the nervous system, it causes symptoms like seizures, tremors, and hallucinations; it also causes blood vessels to constrict, affecting blood flow, and alters the physiology of smooth muscle, leading to such complications as miscarriage and gangrene as well as a variety of gastrointestinal and respiratory symptoms. Fatalities are not uncommon in ergot poisoning, and in some "outbreaks" the number of victims has been significant.[41] Historical records show that arguably the worst occurred in 994 AD in France, where between twenty thousand and forty thousand people died of ergot poisoning.[42]

Despite the all too evident dangers of accidental ergot consumption, ergot nonetheless found its way into our medicine cabinets, used in traditional medicine for a variety of reproductive health issues like inducing childbirth or stopping postpartum bleeding (acting a bit as Pitocin does today). It was also used to treat ailments like migraines, and once specific alkaloids were able to be isolated from it, these were used against hypertension and other maladies.[43]

Ergot is yet another example of the fine line between medicine and poison, but as with psychedelics in general, there has also been much speculation on its use for recreational purposes, and on the role of these chemicals in the development of human cultures. The use of many mind-altering substances has its origin with seeking religious and spiritual experiences, and some thinkers link ergot consumption with a variety of religious rituals. One such notion is the "ergotized beer theory" (which would more correctly be termed a hypothesis). Formally articulated in 1978, it posits that beer prepared using ergot-contaminated grain was used in religious ceremonies—notably the famous Eleusian Mysteries of Ancient Greece*—likely because of its possible hallucinogenic properties. The perception of this idea by scholars alternates between interest and derision depending on the context and the particular scholar. As far as I know, there is no direct evidence in favor of this hypothesis; there do not seem to be any explicit historical records concerning the rationale of this practice.[44]

---

* The Eleusian Mysteries are an appropriately named initiation ritual which, while shrouded in secrecy as to the details, appears in the writings of everyone who was anyone in Ancient Greece, from Aristotle to Homer to Plato, and is a truly fascinating topic in its own right (rite?): https://doi.org/10.1093/acrefore/9780199381135.013.8127.

And while we know of no *nonhuman* animal having even a rudimentary notion of spirituality, with the possible (and only possible) exception of certain intriguing behaviors in a particular troop of chimpanzees,[45] our maverick psychopharmacologist Dr. Ronald K. Siegel (whom we met in chapter one) has even mused[46] about religious-like behavior in certain animals in light of exposure to psychoactive substances—a highly speculative topic that is beyond the scope of this book. This does lead us, however, to yet another "out there" but interesting idea.

## The Stoned Ape Hypothesis[47]

Humans have consumed psychoactive substances for millennia; eventually, psychedelic substances gained a place of honor in the human pharmacopeia,[48] due in no small part to the fact that they induce altered states of awareness virtually indistinguishable from spiritual and mystical experiences. In fact, some scholars give serious thought to the idea that religion and spirituality themselves may have psychopharmacological origins. And at least one scholar believes that the exposure of our ancient ancestor apes to psychedelic substances may have triggered our cognitive development into "true humans." This is the controversial "stoned ape hypothesis," which aspired to explain how humans developed consciousness (as you might imagine, the name alone made a significant contribution to the controversy).[49]

The stoned ape hypothesis (SAH) was an unconventional idea and the brainchild of an unconventional scholar: Terence McKenna was an ethnobotanist/anthropologist by training, and an adventurer at heart.* Perhaps this is not so surprising. Because of their complex roles in human societies, psychedelic substances offer an excellent

---

* Just to give you an idea of the kind of thinker he was, just look at this sentence from his obituary: "Mr. McKenna combined a leprechaun's wit with a poet's sensibility to brew a New Age stew with ingredients including flying saucers, elves, and the *I Ching*" (from nytimes.com/2000/04/09/us/terence-mckenna-53 -dies-patron-of-psychedelic-drugs.html). His story is richer than his hypothesis can tell; for more about him, please see inverse.com/archive/july/2017/science, and samwoolfe.com/2013/05/terence-mckennas-stoned-ape-theory.html.

example of the benefits of looking beyond the traditional purviews of the natural sciences; to fully understand these substances in the context of human affairs, one must consider insights from such disciplines as anthropology, sociology, psychology, and even theology, among others.

McKenna's hypothesis rests upon the foundation of one uncontestable fact: the primate lineage that eventually led to us humans experienced a series of truly puzzling evolutionary phenomena in the realm of cognition—various events of uncertain origin that led to significant increases in the size and eventually the abilities of the brain, changes that were seemingly faster than what the "normal" pace of evolution would predict or even allow. For example, paleontological evidence suggests that the cranial capacity of our early ancestor *Homo erectus* doubled in a relatively short period of time (thousands, as opposed to millions or even billions of years). This increase might have taken place as early as two million years ago but no later than seven hundred thousand years ago.[50] A second event inferred by paleontological data suggests that the brain capacity of *Homo sapiens* *tripled* between five hundred thousand and one hundred thousand years ago.[51] Furthermore, even though the "proper" lineage of modern humans is between two hundred thousand and three hundred thousand years old, it was not until about sixty thousand years ago that the development and evolution of relatively sophisticated culture and technology really took off. Over time, scholars have proposed various factors, for example the evolution and selection of specific genes, that might correlate the increase in brain size and complexity with the concurrent increase in the complexity of hominid societies—complexity that includes such advances as the "taming" of fire and the development of agriculture[52]—but there is no arguing that it remains somewhat mysterious. In 1992, McKenna entered the fray with the stoned ape hypothesis, published in his book *Food of the Gods*.[53]

McKenna proposed a hypothetical "Ur-plant" that could have conferred our early ancestors with an "expanded mind" by consuming it. He methodically considered the possible hallucinogenic sub-stances that early primates would have been exposed to and the possible plants in which these substances would be found, while offering his corollary rationale along the way. For example, the plant must be from Africa, the undeniable cradle of humanity. What's more, since our ancestors lived, thrived, and evolved in the grass-lands, our plant candidate would have to be found in this particular habitat—but distributed throughout a large geographical area, as early primates within our lineage would have been nomadic. Partly because of our exploratory tendencies, this plant must also have been edible "on the go," meaning that it should not have required preparation, mixing with other agents, or cooking, since these cul-tural practices appeared much later in our history—and at any rate, if you lived in an area chock-full of big predators, you couldn't afford to stay in one place for too long, steeping teas and the like. Finally, in order to be sought by early peoples, the plant's effect should have been both desirable and immediately apparent upon consumption (so that they could be correctly attributed to their source). These aforementioned requirements drastically limited the possible plant candidates, and amusingly, after the proverbial "careful consider-ation," it turned out that the most likely candidate was not a plant after all, but hallucinogenic fungi, specifically hallucinogenic fungi that grew on . . . dung.*

So the essence of the stoned ape hypothesis is that by consuming mushrooms similar to modern versions of "magic mushrooms," our early ancestors' brains were exposed to psychedelics similar to psilo-cybin, the main psychoactive chemical in "magic mushrooms." And, somehow, the exposure to psilocybin and related compounds "opened their minds," making their brains more amenable to evolutionary changes leading to the development of consciousness.

---

* Let the jokes begin . . .

This hypothesis was not taken seriously by the vast majority of the scientific community, a state of affairs that persists to this day.[*] However, some scientists, including yours truly, believe that this idea has some merit, if only as a map to guide our thoughts about the mechanism through which psychoactive substances affect the human mind. After all, a formal definition of consciousness does not yet exist, and yet we aspire to understand how chemicals can influence consciousness. I have heard somewhere that one should not ever try to explain an unknown with an unknown, but in this particular affair, it is all we can do for now. It is important to point out, though, that there is no evidence of any kind that supports McKenna's hypothesis. For example, we do not have the foggiest idea of how a psychoactive substance would cause changes in brain physiology that could become genetically transmittable to future generations. This notion is too close to Lamarckism[†] for the comfort of evolutionary biologists, and McKenna offered a few ideas in his book to preemptively counteract this particular anticipated objection to the SAH. In short, the objection can be boiled down to the following question: How did the pursuit of the chemicals produced by these mushrooms get encoded in the human genome? After all, there is no straightforward way for natural selection to cause this to happen. McKenna's solution was to change the defining parameters. In his words:

> The short answer to this objection, one that requires no defense of Lamarck's ideas, is that the presence of psilocybin in the hominid diet changed the parameters of the process of natural selection by changing the

---

* Recently there have been some inklings of support for it, but this is more a willingness to look at it more closely as an interesting educated guess rather than any scientific examination. Alas, I have not heard of an established, well-articulated experimental approach to test this idea.
† In short, the inheritance of acquired characteristics, a notion that is forcefully contested (with good reason) by current evolutionary thinking. However, certain aspects of this idea are being currently taken into consideration by biologists (please see Ward, 2018).

*behavioral patterns upon which that selection was operating. Experimentation with many types of foods was causing a general increase in the numbers of random mutations being offered up to the process of natural selection, while the augmentation of visual acuity, language use, and ritual activity through the use of psilocybin represented new behaviors. One of these new behaviors, language use, previously only a marginally important trait, was suddenly very useful in the context of new hunting and gathering lifestyles. Hence psilocybin inclusion in the diet shifted the parameters of human behavior in favor of patterns of activity that promoted increased language; acquisition of language led to more vocabulary and an expanded memory capacity. The psilocybin-using individuals evolved epigenetic rules or cultural forms that enabled them to survive and reproduce better than other individuals. Eventually the more successful epigenetically based styles of behavior spread through the populations along with the genes that reinforce them. In this fashion the population would evolve genetically and culturally.*

I wouldn't blame you for thinking that this sounds like someone doing a lot of hand waving while speaking very fast. I happen to agree with you, but I still believe that his idea has value. It is unlikely that sufficient evidence will ever be obtained to support the hypothesis, as we cannot repeat the experiment—that is, we do not have a "pre-cognitive-leap" human to treat with psychedelics and watch evolve (not to mention that this is not how evolution works). That being said, it is undeniable that characterizing the SAH as a "half-baked" notion is unjust, as it is clear that McKenna crafted his hypothesis very carefully.

## SMOKE FROM THE GODS

In *Phantastica*, the ambitious cataloging of drugs we discussed briefly at the beginning of this chapter, Dr. Louis Lewin narrates Christopher

Columbus's arrival at Guanahani Island (now San Salvador) in October 1492. Lewin based his account on Columbus's letters, which were translated and published by Fray Bartolomé de las Casas, an early settler and historian of the Americas.* Columbus wrote:

> In the middle of the gulf between the island of Santa María [probably today's Rum Cay Island] and Fernandina [probably today's Long Island] I found a man in a canoe carrying a little piece of bread about as large as the fist, and a gourd of water, and a bit of reddish earth reduced to dust and then kneaded, and some dry leaves which must be a thing very much appreciated among them, because they had already brought me some as a present at San Salvador.

It is widely assumed, and with good reason, that the dry leaves were tobacco leaves. In his book *Historia de las Indias*[54] ("Indias" was an early name for America, as the Spaniards thought they had arrived in India, a long, well-known story), de las Casas describes:

> [M]any people, men and women, going to and from their villages, and always the men with a brand in their hands and certain herbs in order to take their smokes, which are some dry herbs put in a certain leaf, also dry, in the manner of a musket formed of paper, like those the boys make at Eastertide. Having lighted one end of it, by the other they suck, absorb or receive that smoke inside with their breath by which they become benumbed and almost drunk, and so it is said they do not feel fatigue. These muskets as we will call them, they [the native Indians] call tabaco.

---

* A short time after his arrival in the Americas, horrified by the treatment of native peoples by the Spaniards, he advocated against their slavery (ironically, in favor of the slavery of Africans), and later advocated passionately against slavery altogether. A prolific historian who went on to become a priest, he had a very interesting life.

Sound familiar? He continues:

*I knew Spaniards on this island of Espaniola [today's Dominican Republic/Haiti] who were accustomed to take it, and being reprimanded by telling them it was a vice, they made reply that they were unable to cease from using it. I know not what relish or benefit they found therein.*

This is probably one of the earliest accounts of addiction to tobacco, and hence, to nicotine.

## Birds That Bug-Proof Their Nests

Of course, not everyone is a fan of tobacco, and those who turn their noses up at it (metaphorically speaking) include some common garden pests. As we've seen already and will discuss further later, producing chemicals is a strategy many plants use to keep themselves from being eaten by bugs, and clever gardeners have made use of this as well. Many commercial pesticides contain nicotine, and a cursory web search will deliver many recipes for DIY solutions of tobacco water, as well as versions using coffee to take advantage of caffeine's similar pesticidal properties. The procedure is straightforward: Mix the material (either tobacco from cigarettes or coffee grounds) with water and let it soak, typically for at least twenty-four hours. After that, simply filter the solids away and spray the remaining water over your garden. This seems to be very effective, and interestingly, certain birds use a similar approach to bug-proof their homes.

Like virtually every other class of organisms, bird species have evolved a variety of strategies to combat parasites.[55] In 2012, members of the research group of Dr. Monserrat Suárez-Rodríguez at the Universidad Nacional Autónoma de Mexico proposed that individuals of two bird species, house sparrows (*Passer domesticus*) and house finches (*Carpodacus mexicanus*), both use tobacco to combat ectoparasite infections of their nests by mites and ticks. The clue that first sparked this idea for these scientists was finding significant numbers of cigarette butts in the birds' nests.[56] The two main reasons birds would use cigarette butts in their nests are that 1) the butts provided good structural material for the construction of the nests, and 2) nicotine acts as a pesticide. Further research in which the researchers compared the number of ticks and mites in nests with or without material from cigarette butts confirmed that the latter were significantly more "infested."[57] Even though nicotine and other tobacco-related compounds caused genotoxicity (an increased rate of mutations) in the birds' offspring, this material also increased the strength of their immune response and overall survival rate.[58] This is a nice illustration of an instance when a compound may be toxic but provides a survival advantage in such a way that the benefits outweigh the risks, just like any self-respecting medication. These observations, among many other similar examples we'll see in later chapters, show that the innate pharmacological sense of our nonhuman cousins can surprise even the most seasoned scholar.

Now let's take a closer look at a particular aspect of that pharmacological sense: the medicinal use of drugs by sometimes furry pharmacists.

## CHAPTER 3

# NIBBLE ON THIS BLADE OF GRASS AND CALL ME IN THE MORNING

*[A]nimals who are very sick will stop eating everything
and focus on one particular plant species.*
—DR. ELOY RODRÍGUEZ, *A CONVERSATION WITH . . .
ELOY RODRÍGUEZ*

*It may be simply coincidence that the flowers found with
Shanidar IV have medicinal or economic value (at least in
our present knowledge), but the coincidence does raise
speculation about the extent of human spirit in Neanderthals.*
—DR. RALPH S. SOLECKI, *SHANIDAR IV, A
NEANDERTHAL FLOWER BURIAL IN NORTHERN IRAQ*

The Zagros Mountains are part of a large mountain range shared by modern Iran, Turkey, and Iraq. In one of the mountains on Iraq's side, there is a vast cave showing evidence of more or less uninterrupted human habitation going back about one hundred thousand years. This cave is well known by the local inhabitants of the area, who call it "the Big Cave of Shanidar" or, more often nowadays, "Shanidar's cave." To get some perspective on how long humans called Shanidar's cave home, we can think in terms of generations. If we designate a generation as a period of thirty years, this means that roughly three thousand generations of humans lived and died in Shanidar's cave. How many stories of love, hate, generosity, cruelty, happiness, and sadness the walls of this cave must have witnessed! Alas, no written or even oral records of these stories survive. Recorded history only goes back about six thousand years, give or take, which translates to a mere two hundred generations or so, or less than 10 percent of the time that Shanidar's cave was inhabited. Fortunately, thanks to an extensive (and still ongoing) body of research, we have an indirect window into the lives of a certain population of peoples who lived in this cave about sixty thousand years ago.

The first European exploration of Shanidar's cave was in the late 1920s, but the earliest systematic studies of the site began in the 1950s, directed by Dr. Ralph S. Solecki, an archaeologist then affiliated with the University of Michigan. This cave soon proved to be an invaluable source of insights into a now-extinct type of humans, the Neanderthals (*Homo sapiens neanderthalensis* or simply *Homo neanderthalensis*), who lived roughly between 500,000 and 25,000 years ago in many parts of modern Europe, the Middle East, and Asia.[1] Though the Neanderthals as a species no longer exist, some of their genes live on; in fact, recent estimates suggest that about 20 percent of the Neanderthal genome survives in modern humans, otherwise known as *Homo sapiens*. (Can we say, then, that the Neanderthals are only about 80 percent extinct?)* Until recently, scientists thought that only non-African populations had Neanderthal genes, but we now know that is not the case.[2] I need hardly tell you that we humans are quite . . . "social" (if you know what I mean). In addition to "socializing" with Neanderthals, there is strong evidence

---

* Alas, no, I am being silly here—a species is either extant (existing) or extinct (gone).

indicating that *Homo sapiens* interbred with still another human group, the mysterious Denisovans, and with at least one other as yet unidentified species in Eurasia, nicknamed the "superarchaics."[3] For their part, we know that Neanderthals successfully mated both with modern humans as well as other hominin species.* Our family tree grows lusher all the time! In fact, 2019 saw the discovery of yet another member of the human family, *Homo luzonensis*. Who says there is no future in the study of history?

But I digress.

Anyhow, for some time after their discovery in the 1800s, Neanderthals were thought of as the prototypical "primitive" human; in other words, people who hunted, gathered, bred, slept, and . . . did not do much else. Modern archaeology and paleontology have given us a more nuanced understanding of these long-lost cousins. First, it is a well-known fact that, on average, the Neanderthal brain was slightly bigger than the brain of our own species, which suggests—albeit circumstantially—that they would have possessed cognitive capacities at least similar to modern humans. Other evidence supports this: we now know that the Neanderthals made relatively sophisticated tools, and there is some suggestion that they used symbols, giving us a glimpse into their thought processes. Furthermore, based on certain specific anatomical and genetic features, it seems that they were almost certainly capable of speech, and it is not far-fetched to think that they would have developed some language as well. Alas, there is no evidence of Neanderthal written language. However, between 2012 and 2018, archaeologists reported the discovery of at least three caves in modern Spain displaying Neanderthal art. The inhabitants of these caves painted dots, handprints, geometric shapes, and even animals like horses.[4] Additionally, researchers found abundant evidence

---

* *Hominids* include all extant and extinct apes, including us. Please remember that all human species that have ever graced our planet are actually apes. *Hominins* are unmistakably human species as defined by considering them closer to us than chimpanzees and bonobos, our closest living relatives.

of other forms of art, represented by bodily ornaments (shell necklaces, body paint, and so on). Scientists are certain that this art was produced by Neanderthals, since, according to radiometric dating, the oldest site is about 65,000 years old,[5] and current archaeological evidence indicates that *Homo sapiens* did not get to Europe until about 20,000 years ago. In fact, other excavations throughout the Iberian peninsula (essentially Portugal and Spain) have discovered Neanderthal artifacts dated as being close to 115,000 years old. In other words, multiple strands of available evidence suggest that Neanderthals developed at least the beginnings of culture. They were, without a doubt, more human than many would like to believe. And to return to the topic of Shanidar, the study of the remains of eight adult Neanderthal bodies found in one of the cave explorations gave us other intriguing hints about possible Neanderthal practices, including some that have special relevance to our conversation.

As the story goes, shortly after the discovery of the Shanidar remains, Dr. Solecki sent some samples of the ground surrounding the bodies to Dr. Arlette Leroi-Gourhan, a palynologist (an expert in plant pollen) based at the Musée de l'Homme in Paris, France. Dr. Leroi-Gourhan found a significant amount of pollen around the body of Shanidar IV, the remains of a forty- to fifty-year-old man. (As is the standard in archaeological practice, these bodies were given numerical designations and are formally called Shanidar I–VIII.) Analysis of the pollen samples showed that they represented some thirty plant species including yarrow, yellow star-thistle, ephedra, grape hyacinth, and hibiscus, among others. Significantly, the pollen tended to be found in "lumps" as opposed to scattered throughout the ground samples. This distribution pattern was consistent with the idea that the pollen may have come from clusters of flowers purposefully placed on the ground.

Also intriguing were the specific plant species that adorned Shanidar IV's grave. Scholars could tell that the burial took place between May and June, since this is when these plants are in bloom in the geographical vicinity of Shanidar's cave, but this is not the point of most interest for our purposes. Far more significant (for us) is that the types of pollen present, not just the distribution pattern, argued for purposeful placement of these plants. Two main facts supported this argument. The first is that many of the plants produce colorful flowers. It is easy to imagine the man's companions lovingly laying his body down on a beautiful bed of flowers in preparation for the burial itself; we perform variations of this ceremony to this day. But the second fact—and where things get exciting for us—suggests the choice of plants at this burial site might even have gone beyond aesthetics or spiritual meaning: many of the plant species represented at the Shanidar IV burial site have well-known medicinal properties.[6]

Taken together, these findings led scientists to believe that Shanidar IV was one of the earliest examples of a *shaman*, or sacred man, buried in a tomb adorned by the plants he knew, used, and loved. This is admittedly a romantic notion, and although plausible and even reasonable, there is no universal consensus about whether the Shanidar remains constitute evidence of ritual burial practices among the Neanderthals. As with virtually every scientific discovery, there is room for interpretation, a room that is crowded with advocates, detractors, and the undecided. Some scholars believe that other, more prosaic factors better explain the presence of flowers on Shanidar IV's burial site.* However, the story does not seem to be over. In 2019, a group of archaeologists reported on a recent excavation of Shanidar's

---

* In the late 1990s, Dr. Jeffrey D. Sommer, an archaeologist then at the University of Michigan, conducted a reexamination and reinterpretation of the Shanidar IV flower burial hypothesis and proposed a more pedestrian explanation: rodents. Dr. Sommer's argument is based on the fact that there were abundant signs of rodent activity as well as a good number of rodent skeletal remains at the caves. Zoological and zoogeographical evidence pointed at the specific identity of the rodent in question: *Meriones persicus*, commonly known as the Persian jird. *M. persicus* is a rather endearing, furry-tailed, mouselike creature with a habit of foraging and accumulating seeds and plant material—including some of the plants found at the Shanidar IV site. For now, though, unless compelling evidence tells me otherwise, I am sticking with the burial theory.

cave that uncovered another set of Neanderthal remains showing the
same clustered pollen patterns. Interestingly, in this case, there was
also evidence of regular accumulation of pollen at slightly different
times, suggesting that flowers were deposited over the burial site
more than once, presumably year after year, just as we might repeat-
edly take flowers to the grave site of a loved one.[7]

Despite the disagreement around the meaning of the findings
at Shanidar's cave, there is direct evidence pointing to the medicinal consumption—by Neanderthals—of at least some of the plants
found at Shanidar. DNA analysis of Neanderthal dental remains
from El Sidrón, a cave in northwestern Spain, showed that these
humans actually ate these plants. This evidence of plant consumption
seemed odd, as all available evidence overwhelmingly indicates that
the Neanderthal diet consisted predominantly of meat (sometimes
including cannibalism). These facts became easier to reconcile when
another factor was taken into account: the plants in question almost
invariably have a bitter, unpleasant taste, on account of the medicinal
compounds they produce. Assuming Neanderthal taste buds worked
similarly to ours (after all, they shared most of their genes with us;
it is unlikely that their sensory physiology was fundamentally different), one would think they'd have needed a very good reason to eat
something so unsavory.* Researchers proposed that it was unlikely
that these humans consumed these plants for nutritional purposes;
rather, they would have consumed them as remedies. Granted, there
is no hard evidence supporting this interpretation, but the Shanidar
and El Sidrón sites are not the only evidence of prehistoric medicinal
practices.

In 1991, two German mountaineers were exploring the Hauslab-
joch mountain in the Ötztal Alps, located more or less between Austria
and Italy (there was a long-running minor dispute over which country

---

* As we all know, to this day, we regularly add sweet flavoring to most liquid
medications.

could lay claim to this particular mountain). The mountaineers discovered a frozen human body, which they at first thought belonged
to a fellow climber who'd met with an accident. Upon closer examination, however, it was apparent that any such accident would have
had to have occurred quite some time before—what they'd found was
in fact the mummified corpse of a man who lived approximately five
millennia ago. Analysis revealed that the so-called Iceman—dubbed
Ötzi—was not a Neanderthal but a more "modern" human, and his
discovery gave archaeologists and anthropologists a unique glimpse
into the lives of proto-Europeans.[8]

Among the many contributions of this venerable gentleman was
some additional evidence for the use by our ancestors of naturally
occurring substances as medicines. Ötzi's possessions included a
small quantity of the polypore bracket fungus (*Piptoporous betulinus*),
known today to have medicinal properties, including as an antiseptic,
antifungal, and promoter of wound healing—properties that would
understandably have been useful in Ötzi's adventures.[9] And although
Ötzi is an important and so far unique finding, I have a feeling that
as glaciers continue to melt, we will find similar human remains, with
the promise of learning even more about some of those who preceded
us on this earth.

Ötzi, along with the people of Shanidar and El Sidrón, undoubtedly represents a mere sample of the pharmacological practices of
our ancestors. The specific example of possible Neanderthal use of
medicinal plants at Shanidar dates from 60,000 years ago, but these
peoples were around long before that. Furthermore, our own species
is about 300,000 years young. The practical reality is that modern
science works on written published knowledge—as we discussed in
a previous chapter, unless I specify otherwise, when I say "The first
example of X . . ." what I really mean is "The first *reported* example
of X . . ."—but the very first instance of purposeful use of medicinal
plants was never recorded in history. It is certain that long before
recorded history began, our ancestors across the globe knew of

animals who behaved differently when sick, namely by eating certain plants and other materials, and it is certain that at some point these ancestors began engaging in medicinal practices of their own. And regardless of whether these ancestors were "almost humans" (like *Homo erectus* or *Homo habilis*), Neanderthals, *Homo sapiens* like us, or some other human species we have yet to discover, on that fine day, we began walking a path that we are still walking today, with various types of drugs as our faithful companions—a path well-trodden by the animals who lived before us.

## TO KNOW THE MINDS OF OTHER ANIMALS

The question of whether or not our animal cousins possess consciousness has sparked intense debate from the proverbial philosopher's cave to the science lab to the living room. Everybody and their brother thinks (some will categorically say they *know*) that they are conscious, but a precise definition of what being conscious even means has proved irksomely elusive through the ages. Indeed, the exact nature of consciousness is probably both one of the most common and one of the thorniest questions that humans ponder.[10] Proposed answers range from the assertion that consciousness is nothing save firing neurons, with even free will an illusion, to the belief that it is a fundamental phenomenon that is somehow beyond purely physical causes. Perhaps more to the point, regardless of what gives rise to our experience of awareness, there is disagreement as to what amount or kind of awareness is enough to constitute "consciousness."

Personally, I am of two minds—pun absolutely intended—about characterizing most animals as "conscious" beings, which becomes especially relevant as we dive further into the admittedly fascinating behaviors of self-medicating and drug-related "recreational" activities. We have a tendency to ascribe humanlike knowledge and/or thought processes to animals, but this tendency can easily lead us astray. No animal possesses formal chemical knowledge about any plant that it may eat, or about the cause of the experiences or relief that it might experience as a result (though of course neither did we, early on). Moreover, in many cases, these behaviors are instinctual, without evidence of any actual conscious intention on the part of a sick animal to eat the aforementioned plant in order to "feel better." And, of course, there is much we do not know about to what extent various animals are capable of forming such intentional, reasoning-based thoughts in the first place.

On the other hand, it is evident that many animals display behaviors that at least circumstantially hint at cognitive processes, and it is not so far-fetched to think that they would retain some knowledge of beneficial or desirable plants. This seems especially likely in the case of primates. For example, there is evidence that chimpanzees engage in *cumulative culture*. This is a practice through which knowledge is purposefully passed to the next generation—a representative example is chimpanzee parents teaching their offspring to use tools. There are even cultural differences in this practice between various chimpanzee groups—that is, different tribes of chimpanzees "specialize" in certain tools and techniques, as well as vary in the approach they take to teach such practices to their offspring.[11] If you ask me, behaviors like this are not so far from teaching your child that eating a certain plant might help you when you are sick.

Thus, given what we know today, I buy the idea of cognitive processes leading to the intention of eating a plant to alleviate aches and pains in the case of mammals and birds, and I would even give reptiles and amphibians the benefit of the doubt. Yet while it cannot be denied that "lower"* animals (like insects—which we'll discuss

---

* It is important to explain an essential concept in evolutionary theory: It makes no biological sense to say that an organism is more "advanced" than another organism without context attached to the statement. As I said in my book *The First Brain*: "There is this common misconception about the way we describe organisms that implies that some are 'better' or 'more advanced' than

more later) engage extensively in proto-pharmaceutical practices, I am not quite prepared to consider the idea of invertebrates like these engaging in "conscious" medicinal behaviors. That said, some invertebrates like squid, octopi, and so on are undeniably intelligent, and I would not be in the least surprised if further research shows us that octopi engage in (for lack of a better term) marine pharmacology.* Then there are "superorganisms," like insect societies, which collectively display quite sophisticated behaviors. They even use some proto-microbiological knowledge in their agricultural practices—which differ very little from "true," as in human, agricultural practices.[12] So I would not be terribly surprised to learn that an ant colony knows its pharmacology either.

The fact is that many, if not all, animal species display at least some degree of what we call—subjectively—the *phenomenon of mind*. The reason this is a subjective assertion is that most of the evidence we use to infer the presence of a mind in our fellow creatures is circumstantial. After all, strictly speaking, I cannot even be sure that the emotions I feel or the information I perceive from the environment are experienced in the same way by another human. To compare emotions and experiences through the species barrier is orders of magnitude more difficult. With very few exceptions, such as the case of several apes who learned the rudiments of sign language, we lack an unambiguous window into the mental processes of any animal species. And as hard as it is to imagine how a gorilla feels, it is even harder to imagine the feelings of animals that, in many cases, perceive the world in radically different ways than we do.

---

others. We must not confuse 'more complex' with 'advanced' or 'better adapted.' Evolution has no directionality. An organism can be simple in structural terms, yet perfectly adapted to its environment [...] organisms are 'fit' or 'adapted' invariably in the context of the environment in which they live." In this case, our discussion of "higher" and "lower" organisms refers to specific cognitive abilities.
* We will see some octopi on ecstasy in chapter five . . .

Every single organism on this planet has the capacity to detect environmental signals and, based on those signals, act upon the environment's message. We humans, for example, detect and decode our environment through the famous five senses. Our interaction with the world does not go much beyond seeing it, hearing it, smelling it, touching it, or tasting it, and the ranges at which our senses perceive our surroundings have notable limits.* Our visual and hearing systems are able to detect a mere sliver of the wide spectrum of sound and light frequencies, and our tactile capabilities are similarly limited—as are our chemical detection systems, namely smell and taste. It is common knowledge that many organisms display far more sensitive sensory systems than ours. Take the mantis shrimp, which can perceive close to one hundred million colors (the average human can perceive close to one million); as you can imagine, we cannot possibly have an accurate idea of how a mantis shrimp experiences the world.[13] We cannot form a visual image using infrared light (like the predator in the movie franchise of the same name), either, yet many types of vipers, including rattlesnakes, can pull off this trick thanks to specialized organs. Let's not forget bats and dolphins, who perceive their world via echolocation in two different mediums (air and water). I cannot even begin to imagine what it would be like to find my way from place to place through echolocation.[14]

As you can see, we are not equipped to truly experience what it is like to be "the other" in the case of animals (and have even less hope of knowing what it is like to be a redwood[†]). This point was very well put by Dr. Thomas Nagel in the *Philosophical Review* in 1974, in an insightful essay appropriately titled "What Is It Like to Be a Bat?" Dr. Nagel writes: "Even without the benefit of philosophical reflection, anyone who has spent some time in an enclosed space with an excited bat knows what it is to encounter a fundamentally alien form

---

* Although we have learned to dramatically enhance our perception through science and technology.

† We will soon talk about how plants are aware of their environment as well.

of life."[15] We are thus at least somewhat limited in our ability to tell whether and how these "others" perceive pain and suffering—which has no little relevance to the idea of intentional self-medication, as you might expect. This being said, I see no reason why other organisms would be incapable of experiencing both pleasant and unpleasant sensations in a way at least somewhat similar to us. Most scientists (and nonscientists) agree that animals likely *do* experience these sensations and their associated feelings.[16] However, I must note that this consensus applies mostly to our fellow vertebrates; when it comes to the types of animals we collectively call "bugs," the consensus is not as firm. But if you ask me (and you are *reading* me, after all), the notion of both pain and the more subjective experience of suffering is not so far-fetched even in (multicellular) invertebrates. After all, vertebrates and invertebrates share many fundamental aspects of the physiology, biochemistry, genetics, and anatomy of their nervous systems, and it stands to reason that there is at least the potential for these commonalities to generate similar results. This is a very active line of research, and I am sure that this subfield will develop further insights as time goes by.[17]

The question of animal minds in general is as fascinating as you might think, and has been explored masterfully by scientists coming from a variety of disciplines and points of view. Of course, I am confident that if you have ever had a pet, especially a dog or a cat, the question of whether they *have* a mind and emotions is not a question at all. Personally, I am with Darwin when he presciently stated that the "difference in mind between man and the higher animals, great as it is, certainly is one of degree and not of kind." And, as you read along, know that when we talk about frustrated flies, angry bees, or

---

* A related personal anecdote: About two years ago, a bat found its way into our basement, unannounced, and—appearing in my presence unexpectedly—it startled me to a degree that resulted in *"I will never live this down"* levels of amusement for my wife and children.

hallucinating slugs, yes, we are anthropomorphizing, but not as much as you might think.[18]

## A FEW WORDS ON ANIMAL-BASED RESEARCH

The practice of "testing" drugs on animals has likely been around since our ancestors first noticed that some plants produced pleasant or useful sensations and some quite the opposite—and that the same plant might lead to both, depending on the amount. Over time, we've continued to test the effects of various substances on animals for a variety of reasons ranging from the desire to determine the safety of compounds intended for human use, to scientific research (in the case of psychoactive compounds, often aimed at studying the addictive or toxic effects of abused drugs), to sheer curiosity. I would like to state that I do believe (well, I *know*) that research on animals is an important, even essential, contributor to the progress of the biomedical sciences—as long as it is done in a humane way and when there are no other viable alternatives (such as cell-based testing, for example). There is no doubt whatsoever that many of us are alive today in no small part because of animal research, and the animals I speak of include humans, as no medication gains (legal) approval for clinical use without properly conducted human trials. However, the sad reality is that historically there has been very little regard for the humane treatment of experimental animals. A few of the stories in this book relate early research efforts that make this quite clear. This situation has thankfully changed for the better in the last few years, but animal-based research is still a controversial topic.

We are all aware that rats, mice, rabbits, and similar animals are frequently used to test the effect of experimental substances, but this book will not talk much about drug administration to "typical" research animals; rather, we will focus both on stories about what we observe regarding animals encountering drugs on their own as well as what happens when scientists give psychoactive drugs to animals that are not necessarily associated with drug research or that would not naturally encounter a given drug. Most of us are not surprised upon learning that a mammal (say, a rat) reacts to a drug similarly to how humans would; after all, we are mammals, too, and we are used to thinking of rats as "animal models." On the other hand, as we "move away" from the mammalian evolutionary line, our propensity for surprise increases. In general terms, we are most curious about animals that are very different from or very similar to us, as well as those we are the least likely to encounter in our own lives. We (scientists included) get especially curious about psychoactive drug effects in animals, particularly because we know so little about them. For instance, most of us have very little idea of how a garden slug might react to, say, LSD or cocaine.[19] And in some cases, this mystery is both absolute and likely to persist—there is absolutely no information (as far as I know) on how cocaine would affect the behavior of a giant squid because a) there is no natural source of cocaine in the ocean, so our squid friends cannot encounter this substance in the wild, and b) to do behavioral experiments with such a large and relatively scarce (and scary) animal would be impractical at best.

A frequent motivation for giving drugs to animals is, well, to see what happens, but other times it does not even occur to us to think about "what happens" when we give drugs to an animal until circumstances guide our thoughts on these matters. A particular example comes to mind: giving animals anesthetic agents for surgery. Our intent is simply to prevent pain and induce unconsciousness, but veterinarians often observe atypical behaviors in their patients when they come out of anesthesia. Sometimes an unexpected effect comes along, and alert scientists pick up on it and run with it—and by run, I mean experiment.

## PHARMACOGNOSY AND ZOOPHARMACOGNOSY

We've seen that early humans used and consumed a variety of plants for therapeutic purposes well before the systematic recording of historical facts began. Native peoples documented their extensive body of medicinal knowledge through oral traditions, and when things did begin to be written down, knowledge about the medicinal use of plant-derived substances was dutifully captured (the first written accounts date from about six thousand years ago). In the early 1800s, we gave this knowledge a name: *pharmacognosy*, from the Greek words for "drug" and "knowledge." Today, pharmacognosy is defined as "the branch of knowledge concerned with medicinal drugs obtained from plants or other natural sources," or, to use a more formal definition from the American Society of Pharmacognosy, "the study of the physical, chemical, biochemical, and biological properties of drugs, drug substances, or potential drugs or drug substances of natural origin as well as the search for new drugs from natural sources or the study of human use of plant- or mineral-based substances for medicinal purposes."

It is no wonder that at least one author calls pharmacognosy the "oldest modern science"[20]—while in some basic ways these practices have changed little over the millennia, pharmacognosy has taken full advantage of virtually every single one of our technological advances, from the first development of chemical understanding to molecular biology techniques. The original conception of pharmacognosy focused on the use of raw (mostly botanical) materials (for example, certain plant extracts) without regard for the active compounds within these materials that actually caused the desired medicinal effect—largely because such chemical knowledge was not yet available. In time, pharmacognosy not only expanded to include the use of animal- and mineral-based substances, it became a truly systematic field of study that combined our existing knowledge with our growing expertise in the identification, analysis, and study of the physiological properties of relevant compounds.

One important refinement came in the late 1990s, when a group of researchers proposed seeing this discipline as a unit supported by a "tripod" where each leg is a factor shaped by evolutionary forces. The three legs of this tripod were: (1) an organism (i.e., a fungus); (2) its biological activity or activities (i.e., hallucinogenic properties); and (3) the chemical or chemicals responsible for such activities (i.e., ergotamine). And a few years later, fields like chemoinformatics, biotechnology, phylogenetics, systems biology, and even a couple of "omics" (for example, metabolomics and genomics) coalesced with pharmacognosy's "tripod" to further advance and accelerate the discovery of biologically active substances.[21] Technical terms aside, the point is that, these days, we use every scientific tool at our disposal to uncover possibly interesting or useful compounds in nature. But as pharmacognosy began millennia before the origin of systematic scientific research, the question arises: How did our ancient ancestors figure out what to take and what to avoid in the first place? The most likely answer, alas, is trial and error, a system with an obvious impediment: when dealing with the natural world, errors can be fatal. In the world of our ancestors, if you did not get quickly away from that big, beautiful cat you met in the forest, you might find yourself the main course of its lunch. That some dangers were less self-evident made them no less deadly: if you ate the wrong berry, nibbled on the wrong leaf, or munched on the wrong mushroom, you would die as surely as if there were fangs and teeth involved. Pharmacognosy was built on the accumulated knowledge of tragedy, on the lives that our fellow humans lost when they tried the wrong thing at the wrong time or in the wrong amount.* Of course, trial and error isn't all bad—we learned about nature's beneficial entities in the same way.

That animals were eager and sometimes savvy pharmaceutical practitioners before us is no real surprise, as virtually all of the animal species that engage in this practice had a headway of thousands, perhaps even millions, of years before our species was even the spark of a dream in Mother Nature's eye. It is safe to assume that our animal cousins "learned"† what was good and what was bad to eat in much the same way we did, albeit via a slightly different mechanism: the process of *natural selection*. Again, it is unlikely that reasoning played a role in the eventual "adoption" by a species of a particular custom to eat a certain plant to alleviate a malady or in the development of the behavior of avoiding a poisonous plant. Instead (and this explanation is much simplified), if an animal naturally expressed a behavior that made it nibble on a toxic berry, said animal probably died, preventing the passing of its toxic-berry-nibbling genes to the next generation. Conversely, if another animal (say, a dog) ate a beneficial plant (say, some grass) to alleviate an upset stomach, that animal might go on to survive and reproduce. Although there were no apparent conscious processes leading to these behaviors, there is no doubt that they got the job done.

It is important to point out that even once we'd gotten past trial and error, our own early medicinal practices were not strictly rational,

---

* In the words of a social media friend, Ms. Emily Grace Buck: "Do you ever think about how many brilliantly inquisitive early humans died as our species tried to figure out what plants and animals we can and cannot eat?"
† A relatively minor controversy in scientific circles is about whether animal self-medication is learned or instinctual (if you ask me, it is likely to be a combination of the two), but the important thing so far as we are concerned in this book is that there is no question that the phenomenon is real.

either—not in the least. Medicine began with magical and eventually religious beliefs that were based, just like the behaviors of our animal cousins, on previous experience, learning, and instinct. It was only with the early beginnings of science that we began to practice medicine in a more rational way, but there is little doubt that we noticed what our fellow beings did. For example, people in virtually every society that has ever graced our planet know that dogs and cats, when feeling ill, tend to eat grass or leaves, things that are not part of the animal's usual diet. What your pet is doing in these instances is consuming material that they cannot digest, in order to use it as roughage to "expedite" the digestive process, or alternatively, to take advantage of any medicinal properties of the consumed plant.

Since animal self-medication is so widespread, as evidenced in many anecdotal and scientifically documented reports of this practice, you'd be forgiven for thinking that scientists have had a name for this activity since time immemorial. Not so, unless by "time immemorial" you mean "1993," which is when the biochemist/plant biologist Eloy Rodríguez and the primatologist Richard Wrangham coined the term *zoopharmacognosy*. This word means exactly what it seems like: *the study of self-medication behavior in animals.*[22] (An important note: we humans are part of the animal kingdom, so strictly speaking, pharmacognosy is a subset of zoopharmacognosy.)

The birth of zoopharmacognosy as an established scholarly field is an interesting story. According to Dr. Rodríguez,[23] he was having a conversation with Dr. Wrangham—who, as a primatologist working in the Kibale forest in Uganda, was very familiar with the feeding behavior of monkeys, chimps, and other primates, including their feeding behavior when dealing with parasite infections. Dr. Wrangham mentioned what he seemed to regard as somewhat trivial observations, namely that monkeys would eat pretty much any type of plant (including some that displayed evident toxicity) while chimpanzees were more selective, and that furthermore, there were certain plants that chimpanzees generally ate only when suffering from parasites. At the time, the "common wisdom" interpretation of these observations was that monkeys were more resistant than chimpanzees to natural toxins in their environment and that chimpanzees consumed leaves mainly for roughage in order to physically rid themselves of parasites ("physically" meaning through, er, digestive mechanisms, rather than any particular characteristic of the plants themselves).* The roughage idea did not ring completely true to Dr. Rodríguez, and he proposed a collaboration with Dr. Wrangham to explore the possible reasons for this chimpanzee behavior. Drs. Rodríguez and Wrangham eventually demonstrated that chimpanzees did not eat leaves only as a digestive aid after all.† Rather, they did indeed consume specific plants in correlation with conditions like parasitic infections and related symptoms, like fever. Even more interesting, chimpanzees seemed to have an intuitive sense of "dosages." For example, the amount they ate from these specific plants depended upon how sick they were—the worse the infection, the more leaves consumed, and vice versa. Eventually, these scientists' observations led to hypotheses, and thus, zoopharmacognosy as a formal discipline was born.

In 1997, Dr. Michael A. Huffman, a primatologist based at Kyoto University in Japan, formalized Rodríguez and Wrangham's insights by establishing four criteria to be used in determining whether a plant-eating behavior in an animal is truly medicinally motivated.[24]

1.  The plant that the sick animal eats is not a random plant (meaning that it is a specific kind), and this plant must not constitute a normal component of the animal species' diet.

---

* An alternative explanation was that chimpanzees are undoubtedly more intelligent than monkeys.
† Although it is also true that whenever chimpanzees ate spiny leaves, scientists would find significant numbers of parasitic worms entangled in undigested leaves (and yes, they collected the undigested leaves in exactly the way you might imagine they would).

2.  The eaten plant must not provide the animal with a
    significant amount of calories, vitamins, and so on. (The
    key word being "significant"—strictly speaking, virtually
    anything you eat can provide calories. The question is
    whether those few calories contribute in a meaningful way
    to nourishment; in these cases, the consumption has no
    significant nutritional benefit.)
3.  This activity must be seasonal, meaning that it must occur at
    specific times of the year when parasites are more likely to
    infect the hosts.
4.  Healthy animals from the same group do not eat the plant.

If an observed plant-animal relationship fulfilled the four aforementioned criteria, scientists could be reasonably sure that the plant
in question has medicinal properties.*

It is well documented that pharmacognosy was largely inspired
by the observation of animal behavior, and many animal species even
provided us with clues about what eventually became widely used
substances. Dr. Giorgio Samorini, a distinguished ethnobotanist who
hails from Italy, reported that the nineteenth-century work of Dr.
Paolo Mantegazza (of cocaine fame—remember him from the last
chapter?) explicitly discusses the animal-inspired human consumption of drugs.[25] And an apt (and adorable) illustration of this process
comes to us from the same Dr. Huffman who delineated the zoopharmacognosy criteria above.

---

* As with everything in science, definitions are always being refined/modified/
further developed, and these criteria are no exception. There are cases in nature
that do not strictly adhere to this formulation, and we will see some examples
of these shortly.

## BABU KALUNDE AND THE BABY PORCUPINE

According to Dr. Huffman, at the beginning of the twentieth century a certain Tanzanian village was experiencing an epidemic of a dysentery-like illness. Around the same time, as it happens, the village's medicine man, Babu Kalunde, had adopted a baby porcupine that had lost its mother. One day, Kalunde's young charge got sick with dysentery himself. Kalunde observed that the animal stopped eating, explored for a while, and eventually dug up the root of a certain plant and nibbled on it. Curiously, Kalunde noticed that the young porcupine seemed to get better a short time afterward. The plant was called *mulengelele*—everybody in the village knew about this plant, but up to that point had carefully avoided it on account of its known poisonous nature. However, in the tradition of some of the best biomedical scientists, the good doctor Babu Kalunde decided to test mulengelele as a dysentery remedy . . . on himself (surprise, surprise), by taking small amounts of the plant when he got sick. It seems to have worked, as he eventually got better and told the people of his village—who apparently listened, as the plant is still part of this tribe's pharmacopeia. To this day, the grandson of Babu Kalunde, Mohamedi Seifu Kalunde (who is also a medicine man), uses the mulengelele plant as medicine for a variety of illnesses, including dysentery.[26]

This baby porcupine is only one example of an animal who led us to a medicinal product. Just like the wise Babu Kalunde, ancient peoples identified plants with useful (or undesirable) properties by observing animals undergoing self-medication, and these properties included the ability to kill or purge parasites from the body, to get rid of a fever, to alleviate pain, to get rid of an infection, and, of course, to invite altered states of consciousness. "Formal" science, however, was slow to recognize many animal self-medication practices as such, and while we now know that animals self-medicate for a variety of reasons, the initial observations leading to this idea were nearly all related to parasitic infections—perhaps not so surprising, as infection by parasites is indeed one of the main phenomena triggering animal self-medication behavior. In general, we can classify the factors that trigger self-medication in animals into two main categories: nutritional reasons (in order to acquire compounds not produced by the organism, as when we consume certain fruits for their vitamin C content) and predation (including the predatory activity known as parasitism).

The earliest published study that speculated about possible antiparasitic effects of plants consumed by primates appeared in 1978, authored by Dr. Daniel H. Janzen of the University of Pennsylvania, a few years before zoopharmacognosy got its name.[27] In his paper, Dr. Janzen reviewed the evidence available at the time in a section whimsically titled: "Vertebrates as Doctors."* There he speculated that animal self-medication was a behavior especially likely to be found in parasite-infected animals. However, in a surprising twist, scientists subsequently discovered that certain vertebrates do not only self-medicate when they feel ill; some also take their "medication" when feeling perfectly well, seemingly to avoid getting sick in the first place. In other words, animal species not only pioneered pharmacy practices; they also pioneered preventative medicine.

## EAT YOUR MEDICINE, PLEASE

Along with consuming "beneficial" substances in order to fight or prevent infection and so on, certain organisms consume material in order to acquire toxic chemicals for defensive or other purposes. Collectively, these behaviors are termed *pharmacophagy*.† The earliest

---

* He even whimsically remarked, "I would like to ask if plant-eating vertebrates may do it on occasion as a way of writing their own prescriptions" (Janzen, 1978).
† Literally meaning "remedy eating." It is important to note some confusion around how these terms are used—some use this interchangeably with zoopharmacognosy. Here we will use it in the narrower sense described.

reference to this behavior I was able to find dates from 1892,[28] in a paper by the German physician and entomologist Erich Haase.* Specifically, Dr. Haase discovered a new genus of butterflies and named it *Pharmacophagous*, on account of his observation that the larvae of many species of this genus fed on a particular plant known for its toxic metabolites. The assumption at the time was that somehow these insects took the toxic substances from the plant and used them for their own nutritional needs. As more examples emerged of animals engaging in these types of practices, it became evident that there were different modalities at work; thus, pharmacophagy was ripe for a redefinition. In 1984, Dr. Michael Boppré, a chemical ecologist at the University of Freiburg, proposed a redefinition of insect pharmacophagy that accounted for the fact that it was sometimes not intended as a nutritional activity. Rather, some insects took plant-derived toxic substances and co-opted them to enhance their survival capabilities, either as protection against predators or as substances that rid them of maladies. In his words, "Plants are not only 'grocery stores' but can also be 'pharmacies' which . . . supply chemicals not needed for their primary metabolism but significantly affecting their [evolutionary] fitness."[29]

It is important, of course, to be reminded that self-medication needn't be limited to plant products. Suitable chemicals can come from fungi, microorganisms, and even other animals. Many cases of pharmacophagy involve animals consuming other animals in order to use a chemical they produce. One of my favorite examples is the strange case of certain sea slugs (generically called nudibranchs) who eat certain jellyfish. Arguably the best-known characteristic of most jellyfish species is that they sting. They do so by producing venoms and delivering them via what are essentially microscopic harpoons. Well, the sea slugs in question not only seem to be immune

---

* You will find that in the 1800s it was not uncommon for physicians to have "side careers."

to jellyfish venom, but when they consume a jellyfish, they sequester their prey's microscopic harpoons along with its microscopic venom sack and install the pair in their own skin. That's right: sea slugs don't just eat jellyfish; *they steal their weapons.* Talk about adding insult to injury.[30]

Then again, some medicinal substances do not come from biological sources at all. Consider the practice of *geophagy*, which is the "intentional consumption of earth" (as in clay, rocks, or plain old dirt).[31] Initially considered a simple pathological tendency without any benefit,[32] there is now a large body of evidence indicating that geophagy is not only not pathological (and even in some cases beneficial), but also that it is an ancient human practice dating all the way back to our early ancestor *Homo habilis*, who lived between one and two million years ago. Geophagy, believe it or not, has been a widespread activity at one point or another in virtually every human culture.[33] Moreover, many types of animals, primarily vertebrates, engage in geophagy. Despite being so common, however, geophagy's biological purpose is not completely clear.[34] Some schools of thought hypothesize that it is a nonadaptive trait (meaning that it neither contributes to nor harms chances of survival, like eye color or whether you show dimples when smiling),* and some scholars believe the behavior is triggered simply by extreme hunger. Because there are several maladies that get better when certain minerals are consumed, a competing hypothesis posits that nutritional deficiencies trigger geophagy, while yet another hypothesis proposes that the ingestion of minerals like clay might serve as protection against environmental toxins and parasites.[35] As is often the case in the world of biology, the right answer might well be a combination of these ideas. As of this writing, the jury is still out.

---

* The fact that some people find dimples attractive notwithstanding.

## BUGS THAT USE DRUGS?

Animal-world members of the practical pharmacists' club include parrots, dogs, elephants, and, of course, primates, among many others—we'll hear a few further stories of vertebrate zoopharmacognosy later on in the book.[36] We tend to associate the kinds of behaviors represented by the concept of self-medication with "higher" animals, yet many of the most interesting examples come to us from the world of invertebrates. More likely than not, the phrase "self-medication in insects" seems a bit strange at best, and one would not expect invertebrates to administer medicinal or recreational compounds to other members of their species (or to members of other species, for that matter)—yet these are very real phenomena.*

The fact that "lower" organisms, particularly insects, engage in self-medication practices was only formally recognized in the early 2000s.[37] According to Dr. Jessica Abbott of Sweden's Lund University, behavioral activities such as self-medication are examples of a phenomenon called *adaptive plasticity*, which occurs when an organism engages in particular behaviors in response to a specific environmental challenge—for example, when an insect feeds on a certain plant *only* when infected by a parasite. However, as we saw above, there are behaviors that can be considered as nothing less than examples of preventative medication, such as when an insect consumes a substance that reduces or altogether prevents parasitic infections in the absence of an immediate threat. Remarkably, there are even examples of *transgenerational* medication, in which an insect uses certain substances to protect their eggs or offspring against parasites or other predators. One of the best-studied examples of this is the monarch butterfly (*Danaus plexippus*), that iconic flying creature in orange and black known for their majestic mass migrations.[38] These

---

* Even more surprisingly, drug self-administration in certain insects goes beyond "health" reasons. Some invertebrates consume drugs (for lack of a better word) "recreationally." We will see more about this in further chapters.

insects are found everywhere in the world and have mastered a rather remarkable ability: they are resistant to a series of compounds collectively called *cardenolides* (Figure 3.1).

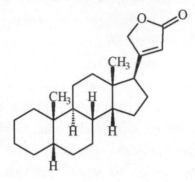

*Figure 3.1. A cardenolide example. Drawn by the author.*

Cardenolides are steroid-derived molecules, and despite their beautiful chemical structure, seen above, all are toxic. They are found in several types of plants, but most commonly in the more than three hundred species of the genus *Asclepias*—milkweeds.[39] Monarch butterflies display resistance to cardenolides by virtue of a mutant version of a specific protein, the sodium-potassium pump, which is critical to the proper functioning of cells, and which, in its nonmutant version, cardenolides wreak havoc upon.[40] Not only are monarchs resistant to cardenolides, they also store them in their bodies at relatively high amounts, for use as a chemical defense against predators. Cardenolides make the butterflies taste bad, usually making the hungry predator spit it out (and perhaps avoid it in the future); should they fail to do so, cardenolides happen to act as a cardiotoxin, making this a mistake a vertebrate predator is unlikely to make twice.[41] Even better, recent work shows that cardenolides negatively affect monarch butterflies' parasites as well.[42]

In addition to the obvious advantage adult monarchs gain by their resistance to and accumulation of cardenolides, the butterflies also take advantage of their chemical mastery to protect their eggs and offspring from predators by preferentially laying their eggs on milkweed plants—but they do not stop there. In a remarkable example of *transgenerational immunity*,[43] monarch parents' habit of feeding on cardenolide-containing milkweed plants makes their eggs resistant to infection by parasites. Interestingly, it seems the eggs display more resistance when the male parent feeds on milkweed plants. The actual mechanism that makes the "father-derived" cardenolides more potent is as of yet unknown.[44] Anyway, the cardenolides that the parents "seed" in the eggs are not completely metabolized; the hatched larvae and eventually the caterpillars themselves retain some cardenolide content, and in turn, the larvae and the caterpillars are chemically protected.[45] It seems the eggs and larvae may be less resistant to cardenolides than adult butterflies, based on recent research showing that the eggs are usually deposited on leaves containing a lower concentration of the steroid. One possible explanation for this behavior is that, since the eggs already have cardenolides in them, if they were deposited in places rich in the chemical, they would "overdose" on it. However, some experiments do not show a straightforward relationship between cardenolide resistance and its presence in eggs—clearly, more research is needed.[46]

The story of the coevolution of milkweed plants and monarch butterflies is more complex and nuanced than what we can explore here.[47] Many aspects of the monarch/milkweed relationship remain unknown, with a number of interesting questions yet to be answered. Regardless of the details, what is certain is that when it comes to cardenolides, monarchs have truly mastered the art of turning lemons into lemonade—and this is a bona fide example of nothing other than preventative medicine, invertebrate style.

## Bee-havioral Immunity

As mentioned in an earlier chapter, nicotine's natural "purpose" is as a pesticide. Not only can nibbling a little too much on a tobacco plant kill virtually any insect, nibbling just a little bit* is often unpleasant enough to "convince" or "nudge" the tiny animal to go find its sustenance elsewhere.[48] However, you will notice I said *virtually* any insect—there are certain insect species that ingest nicotine with relative impunity for reasons that are, for all intents and purposes, medicinal. One of these is the bumblebee.

One of the "occupational hazards" faced by pollinators like bees is that the sorts of flowers they feed upon are also visited by a veritable army of other animals, elevating the risk of contracting infectious diseases, much like drinking water from a contaminated water fountain.[49] *Bombus terrestris*, a European species of bumblebee, is thus oftentimes infected by a variety of parasites, including *Crithidia bombi*, which is similar to the parasite that causes sleeping sickness and Chagas disease in humans. A recent report of the possible medicinal use of nicotine by bumblebees to combat these parasites came from the research group of Dr. Lars Chittka, based at Queen Mary University of London. It offers an example of a type of immunity not mediated by entities like antibodies, but rather dependent on actions—this is appropriately called *behavioral immunity*, and it plays an important role in combating infections in several types of organisms. Dr. Chittka and his collaborators found that when (but only when) they had been infected with *C. bombi*, captive bees preferred nicotine-containing sugar solutions; infection with this parasite also seemed to increase the consumption of nicotine-laced nectar in free bumblebees when foraging. And the nicotine consumed in this way does indeed seem to be effective against *C. bombi* infection in bumblebees. Oddly, nicotine was not found to be toxic to the bees' parasites, which in my mind† is evidence that a nicotine metabolite, rather

---

* The specific amount of any particular substance that qualifies as "a little bit" of course varies widely—from insect to insect, and even depending on the insect's developmental stage.

† Thinking as a pharmacologist, although very interesting, such a mechanism is not at all surprising. There are many examples of drugs that are inert, doing

than nicotine itself, is the chemical that helps bumblebees fight said parasites. This is a possibility that the authors of the study proposed in their conclusions; I expect more research on this will be forthcoming. We know that nicotine is extensively metabolized by bees, reinforcing the possibility that a nicotine metabolite is responsible for the apparent antiparasitic effect against *C. bombi*.[50] Other compounds produced by tobacco plants also seem to have a role to play in this battle of parasite-versus-host. A group led by Dr. L. S. Adler of the University of Massachusetts at Amherst published a series of articles in which they studied *anabasine*, a chemical present in tobacco plants that shows some structural similarities to nicotine, but has slightly different pharmacological effects (Figure 3.2). Dr. Adler's group determined that anabasine also offers bumblebees a measure of protection against *C. bombi*.[51]

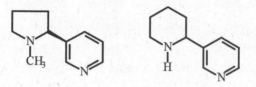

NICOTINE        ANABASINE

*Figure 3.2. Nicotine and anabasine. Drawn by the author.*

Interestingly, experiments show that if healthy (noninfected) bumblebees consistently take nicotine in their food, they get weaker and become sick in other ways; parasite-infected bees tolerate nicotine much better. It seems there is a biochemical nicotine-detoxification mechanism in bumblebees that "activates" upon infection by parasites.

Although the authors of the studies referenced above are properly cautious about "conclusion overreaching," these reports provide

---

absolutely nothing, until they are "activated" by metabolism; these are called *prodrugs*.

tantalizing evidence of yet another example of self-medication in social insects, and suggest that there is much more still to discover. Bees in particular are truly remarkable creatures that express intricate social practices, as well as a variety of curious behaviors when exposed to drugs. We'll see more of both in this chapter and the next one.

## MASTER SMELLERS AND BIOMIMICRY

One of my favorite movies is *Finding Nemo*, a story about a dad, Marlin (who is in fact not a marlin but a type of clownfish, *Amphiprion ocellaris*), who is willing to do pretty much anything to find his lost son (the titular Nemo). Along the way, Marlin teams up with a charming companion, Dory (a royal blue tang, *Paracanthurus hepatus*), who has some of the best lines in the movie. One of these lines—and a favorite for soon-to-be obvious reasons—comes when Dory and Marlin are racing: "Give it up, old man—you can't fight evolution. I was built for speed!"

I wholeheartedly agree with Dory: evolution has had billions of years to tinker with biology, and in the process has produced a veritable world of wonders. Among these wonders are many unique characteristics of various living beings that we humans have noticed and thought might be useful for our own purposes. Throughout history, we've often tried to harness our mastery of science and engineering to imitate these traits. In the 1990s, this practice was given a name: *biomimicry*, the design of engineering solutions inspired by nature (sometimes called *biomimetics*).[52] These efforts implicitly follow Dory's mantra: "you can't fight evolution." Biomimicry is a flourishing area of research and technology, and it has already resulted in quite a few innovative inventions, including the design of novel materials, mechanical parts, and medications, to name a few.[53]

Along similar lines, I am sure you are familiar with the almost magical ability that dogs have to detect smells. Since our early history, we have used sniffing dogs for hunting, to track the whereabouts of missing people, and more recently—and relevant to our topic—to detect the presence of illegal drugs. This development makes sense (ha ha), as the sense of smell is literally a function of detection and identification of chemicals present in air or in water. At the molecular level, it is all about chemical structure: specific receptors in the olfactory system of many species detect these structures,* and once detected, they are identified (perceived as smells). Dogs and related organisms are "master smellers"† due to the expression of specific proteins that can detect hundreds of thousands—in fact, likely more—of distinct chemical signatures. But while dogs get all the glory, there are other (admittedly less cuddly) animals with even more impressive abilities in this area. This honor belongs to many invertebrate species, including certain cockroaches, honeybees, and some 1,600 species of moths. Many moths are adept at detecting chemicals, including pheromones, floating in the air, and they can do so with exquisite sensitivity despite not having proper noses; their chemical sense organs are actually their antennae.[54] Bees, too, are better than most dog breeds—in terms of numbers, their sense of smell is about fifty times more powerful than that of a dog.

Humans have tried to build contraptions that imitate the "smelling" capabilities of animals, with some success, but we

---

* Much as described in the "Philosophy of Pharmacology" discussion in the previous chapter.
† Admittedly, this sounds like an insult you would hear in primary school.

still have a long way to go before we get to the level of our insect friends. After all, insects have been on this planet for almost five hundred million years,[55] which means that evolutionary processes have been at work in these organisms for at least that long. Studies aimed at using insects themselves as living chemical detectors have been going on since the 1960s, spearheaded by the US Army. Insects can be trained to exhibit specific behaviors in response to environmental clues, including odors, and so through the observation of these behaviors, we can infer the presence of chemicals of interest. At first, the main objective of these studies was to detect carbon dioxide and bodily odors that could betray the presence of hidden enemy personnel. Scientists further applied these ideas to the detection of explosives using a variety of insect species. And recently, a scientific paper described the possible application of honeybees as "drug detectors."[56] The authors of this study argued in favor of using bees as a substitute for sniffing dogs for several reasons in addition to their odor-detecting abilities—namely, cost and versatility. In the paper, the authors point out that drug laws change; for example, cannabis is currently legal in many US states but not in others, a fact that proves problematic since in general sniffing dogs are currently trained to "alert" at the presence of any of a variety of abused drugs—it is more costly, timewise, to train a dog to react to cocaine but ignore cannabis. Remarkably, it takes insects like bees a matter of minutes to be trained to distinguish between various drug types, ignoring some odors while reacting to others. We cannot compete with evolution, but why would we want to? The smart thing to do is take advantage of what nature has to teach us.

## WHEN A WHOLE SOCIETY TAKES ITS MEDICINE

While we are accustomed to thinking of human civilizations as the pinnacle of sophistication, many insect species have developed the art of social organization to an exquisite degree. Ant colonies and beehives, among others, belong to a class of highly structured organism societies sometimes described as *superorganisms*. The technical term for superorganisms is *eusocial organisms* (eusocial literally means "truly social"), and animal societies defined as such possess a set of unique characteristics, mostly related to specialization. Eusociality is surprisingly widespread in nature, from insects, to crustaceans, and all the way to certain mammals. Despite the wide variety of the species that express eusociality, there are certain characteristics that define all eusocial societies, namely, (1) the adult population divides itself into at least two main castes: workers (who do not generally reproduce) and a reproductive caste (i.e., queens and drones), (2) a minimum of two generations coexists in the collective, and (3) the worker caste is in charge of eggs and juvenile members of the colony.

One of the characteristics of superorganisms is that the collective interaction between members generates behaviors that go beyond the brain (mind?) capacity of any single individual of that species.[57] Eusocial organisms like ant colonies, then, are "complex minds" made out of many "simple minds." The biology of superorganisms is fascinating, and the specific details of how they run their societies are still being intensively studied.[58]

Perhaps unsurprisingly, many species of eusocial invertebrates even "take their medicine" collectively. An example from the bee world involves the practice of collecting resin material from a variety of plant species and incorporating that resin into the structure of their communal nest. Scientists hypothesize that this behavior helps prevent microbial or fungal infections in the hive. A (fittingly!) collaborative work between Drs. Michael D. Simone-Finstrom and Marla Spivak of North Carolina State University and the University of Minnesota, respectively, further explored this strategy in the common honeybee. Prior to this study, the authors established that the presence of particular resins in bee nests seemed to correlate with a lower incidence of pathogenic bacteria, so they hypothesized that this mechanism might be useful to keep other infections at bay, particularly the fungus *Ascosphaera apis*, a common pathogen in bee colonies.[59] They looked at whether resin collection and incorporation into the beehive affected the presence of *A. apis*, and found that not only did resins decrease the severity of *A. apis* infections, but that bees increased their resin-collecting activities when they detected the pathogen in their midst. This is a particularly remarkable example of animal self-medication because it represents communal self-medication—that is, in true superorganism form, the colony medicates itself. These insects are practicing medicine not just at the individual level, but at the level of their society as a whole. Appropriately, Drs. Spivak and Simone-Finstrom call this type of collective behavior "social medication."[60]

Of all the aspects of our relationship with the chemical entities we call drugs, medication merits a place of honor not just because of the impact it has had on us humans, but because of what it reveals about the resourcefulness of animals more generally in using the materials at hand to make the best of the environments in which they find themselves. Even our invertebrate cousins practice self-medication against infective agents, they protect their offspring using chemical methods (medicines), and perhaps most remarkable of all, they engage in what for all practical purposes are *public health practices*.

As ever, if this book were a song, its refrain would be: *Other animals are not so different from us.*

## CHAPTER 4

# BETTER LIFE THROUGH CHEMISTRY, COURTESY OF PLANTS

*It has always pleased me to exalt plants in
the scale of organized beings.*
—CHARLES ROBERT DARWIN, *THE
AUTOBIOGRAPHY OF CHARLES DARWIN*

*In order to apply appropriately such terms as Mind,
Consciousness, Intention, Design, Desire, to plants, it
is obvious we must change, or, at least enlarge, our
conceptions of their character, and our Definitions.*
—DR. WILLIAM LAUDER LINDSAY, "MIND IN PLANTS"

Imagine that you are taking a stroll in a forest during late spring or early summer. It is a gorgeous day. As you walk, you become aware of the many sounds of nature: the scratchy chirping of cicadas, the buzzing sounds that bees and their ilk make while foraging for nectar, the melodic call of a chickadee, and perhaps the chittering of a pair of squirrels. All of these noises, and indeed the very presence of this crowded soundscape, feel *natural*.

Now see yourself in your mind's eye walking through that same forest, this time completely devoid of animal life. Maybe the only sound you hear, aside from your own footsteps, is the rustle of leaves as they are moved by a light breeze. Oddly, instead of the peaceful sensations you'd expect from a walk through a sun-dappled glade, as you walk, you begin to feel uneasy. The forest is simply too quiet.* And, in fact, this uneasiness makes some sense—our ancestors would have learned the hard way that if the sounds of the forest quiet, especially suddenly, chances are that a predator is on the prowl. But more than that, walking in a forest bereft of animals feels unnatural because it feels . . . *lifeless*. The eerie silence makes you feel as if you are all alone, the only living thing moving through a world gone still.

Alas, things are often not as they appear—in this case, not even close. The undeniable truth is that an animal-less forest is far from inanimate, and as you walk through it, feeling "alone," you are in fact surrounded by living things engaged in a "quiet riot" of activity. Plants, fungi, and related beings† are a perfect example of the popular saying "It is the quiet ones that you have to watch out for," because these "quiet ones" are without a doubt a true force of nature. There is a tendency to think of plants as the background against which life plays out, but plants are anything but passive; they are active players in this drama themselves. In the roughly 1.5 billion years of evolutionary time since plants and animals diverged,[1] plants, just like animals, have evolved an almost countless variety of survival tactics. Virtually all of the psychoactive compounds that we talk about in this book are naturally occurring substances (or versions of said natural products)

---

* You may recognize this phrasing from numerous suspense or horror movies, many of which include some variation of "It is too quiet in here."

† In most cases, any general characteristics that apply to plants also apply to fungi (photosynthesis being one of the main exceptions). For this reason, in order to streamline our story—unless I state otherwise—whenever I refer to "plants," you may assume I mean "plants and fungi." Incidentally, "evolution-wise," fungi are more closely related to animals than to plants. (I *know*! Weird.)

produced by these deceptively peaceful entities with the purpose of subduing an enemy.

## THE BEHAVIOR OF THE UNHURRIED

Seen in the context of our own experience as part of the animal kingdom, the general phenomenon of behavior is easy to understand. You are no doubt familiar with many of the ways—from the usual, to the odd, to the undeniably weird—in which animals respond to changes in their environment. However, the idea of plant behavior is less familiar to us. If asked about it, what usually comes to mind are plants like the Venus flytrap, and other, similar specimens that trap insects and small animals (incidentally, even small vertebrates). However, the phenomenon of plant behavior goes well beyond the obviously carnivorous.

The frame of mind of ascribing plants a supporting role as opposed to seeing them as a central part of life on our planet is long-standing. (Furthermore, this human tendency has a name: *plant blindness*. This is a fascinating topic with many unique interpretations and theories surrounding it, and while it is a bit beyond the proper scope of this chapter, I have discussed it at more length in the endnotes.[2]) It took many years and much research before this attitude began to change, but by the 1800s the scientific consensus about plants began to shift in favor of seeing them as active entities. One of the earliest formal writings expressing this idea was a scientific paper published in 1876, provocatively titled "Mind in Plants." This paper appeared, of all places, in the *Journal of Mental Science* (now the *British Journal of Psychiatry*), and its author was Dr. William Lauder Lindsay, the Scottish physician we met in chapter one in the context of his studies on alcohol and animals (you may remember his "inebriation scale"). One would not expect to find a paper about plants published in a journal with the words "mental" or "psychiatry" in its title. Not only an example of how seemingly disparate subjects can coalesce, this paper also showcases Dr. Lindsay's eclectic intellectual life. He was a formally trained physician with a special interest in psychiatric conditions, a subject that was only beginning to be taken seriously by the medical profession. Remarkably, he was *also* an award-winning botanist, and won accolades for his thesis on lichens (which are at least partially plants). It is not difficult to imagine how his interests came to be integrated. Dr. Lindsay wrote that he was keenly curious about whether the "attributes of mind" were present in both animals and plants, and he engaged in many detailed studies in an attempt to throw light on this problem. "Mind in Plants" included an extensive list of physiological and behavioral characteristics that he thought plants might have in common with animals:

1. *Respiration.*
2. *Circulation.*
3. *Nutrition.*
4. *Digestion—of animal food.*
5. *Secretion: including a solvent juice resembling the gastric.*
6. *Absorption.*
7. *Luminosity.*
8. *Evolution of heat.*
9. *Presence of electric currents.*
10. *Sleep.*
11. *Exhaustion: with reinvigoration after rest.*
12. *Spontaneous movements.*
13. *Same kinds of Diseases.*
14. *Same influence of atmospheric or gaseous Poisons.*
15. *Same results of chemical or mechanical Irritation.*
16. *Same effects of light and darkness, and of heat and cold.*
17. *Contractility—analogous to muscular.*
18. *Heredity.*
19. *Mimicry.*

Lindsay based his insights into the undeniable reality of plant behavior on his own observations as well as those of others—initially, what caught the attention of many of these scholars was the case of carnivorous plants, since they explicitly straddled the worlds of "the hunter" (a predator), and "the hunted" (prey).

In the course of his research, Lindsay surprised himself* by concluding that he could not find an unambiguous line between the "plant mind" and the "animal mind" and neither could he find such a clear frontier between "animal minds" and "human minds." He began "Mind in Plants" with the following:

> *I have encountered as great difficulty in drawing any definite or definable psychical line of demarcation between Plants and the Lowest Animals as between the Higher Animals and Man. In other words, it appears to me that certain attributes of mind, as it occurs in Man, are common to Plants.*

As mentioned, Lindsay was not the only deep thinker with such a daring—at the time—opinion. As quoted in the last chapter, Charles Darwin famously wrote that he believed the difference between human and animal minds to be one of degree (Lindsay, of course, added plants to the club). Indeed, great minds *do* seem to think alike.

One of the best-known defensive behaviors in response to a perceived danger is the *fight or flight response*, which is a phenomenon seemingly exclusive to animals. The first choice is generally "flight," which makes sense, as it is metabolically cheaper to flee or hide than to expend valuable energy fighting—and of course it is also less likely to expose the animal to potentially lethal damage. Therefore, it is usually only when fleeing or hiding is not an option that an animal resorts to fighting. But regardless of what the animal decides to do, it is likely to involve some kind of movement, preferably *fast* movement, as speed

---

* One of the best things that can happen to a curious scientist!

is oftentimes the difference between life and death. That being said, not all animal species have the option of outrunning an enemy; some are very slow or even—as in the case of barnacles and their ilk—fixed in their environment and have therefore had to develop other strategies in order to survive. These include things like sheer size (think elephants and whales),* the capacity to produce strong electrical currents (some eels, among others), the ability to coat oneself with toxic substances (many creatures, among them the African crested rat, who chews the bark of a poisonous plant without any apparent deleterious effects, mixes it with saliva, and then proceeds to coat its fur with it[3]), and so on. These are only a few examples that illustrate the fact that, when it comes to staying alive, speed isn't the only option.[4] It's a good thing, too, as virtually every known species of plant lacks the option of displaying significant motility, as once they sprout in a place, they tend to stay there for life.

A fundamental difference between plants and animals is that plants do not possess an obvious control center or any clearly defined organ systems. This apparent simplicity is in fact an advantage, and one that more than compensates for their relative immobility. Let's think about the two biggest known organisms belonging to the plant and animal kingdoms, respectively: sequoias (redwoods) and blue whales.[5] A well-aimed harpoon can (sadly) kill even the most massive and majestic blue whale. On the other hand, no single blow can kill a redwood (or any other tree, for that matter). If, armed with our biggest, meanest chainsaw, we manage to fell a tree but leave the stump alone, eventually the tree will reassert itself. In other words, there is no single place where we can "hit" a tree and kill it. Although we humans with our tools can get around even this impediment (we are very good at killing things), very few organisms in nature can. And this stubborn resilience—noted by many a gardener—is without

---

* Alas, although very effective against other animals in their environment, sadly, sheer size is no match for human technology.

a doubt one of the main reasons plant life dominates virtually every ecological niche on the surface of our planet.*

Another factor in their incredible proliferation and persistence is their sensory abilities. Plants possess a multiplicity of special senses—some that mirror ours, as well as some that are utterly alien to us—that allow them to perceive and react to whatever the environment throws at them.[6] As far as defensive strategies go, their "decentralized" nature is an effective one, but it admittedly doesn't sound much like the "behavior" we alluded to earlier, yet plants also detect and respond to environmental stimuli and threats, just as animals do. If we use the senses we all know and love as a starting point, we will see some surprising commonalities. For example, plants are capable of detecting selected wavelengths of electromagnetic radiation, a.k.a. light; in other words, they have their own way of "seeing" (this is a process completely independent of photosynthesis). They can detect physical stimuli transmitted by solid materials (touch) or air (sound). They are *quite* adept at detecting chemicals (smell and taste). Additionally, plants have the ability to detect gravitational fields (this is how they tell "up" from "down"), and possess many other cunning talents of perception that we lack.† They use this information to maximize their chances of survival. For instance, there is no "right way" to

---

* In fact, plant life is so ubiquitous that we define discrete land-based ecological communities—called *biomes*—in terms of their plant life.

† Some authors assert that plants use at least fifteen(!) other senses in addition to the "traditional" five. However, I suspect that the actual number of plant senses is likely a matter of our subjective interpretation of how plants see the world. For example, plants are exquisitely adept at detecting the level of humidity in their environment. Some scientists count this ability as an independent sense. On the other hand, we could consider detecting the amount of humidity in the air a special case of chemoreception, pretty much like the sense of smell. This is hardly a controversial statement. To some extent, we are able to detect whether our environment is high or low in humidity, though I do not know of anyone who is able to smell the air and determine the exact percentage of humidity. Even if we decide to count this plant ability as a version of our sense of smell, try for a moment to call up the smell of pure water. Can you? No? Well, neither can I.

plant a seed. What I mean by this is that, try as you might, you cannot plant a seed "upside down." Not only does the tiny plant embryo detect the direction of the gravitational field (thus telling "down" from "up"), it also detects the direction of incident light, and literally moves toward it. By using the special senses that allow them to be acutely aware of their environment, plants can distinguish friend from foe (and sometimes turn foes into friends); they are known to form alliances with other plants and sometimes even with animals![7]

As biologists began to understand plants as "active" organisms, this change of mind (pun absolutely intended) led to some unexpected issues. You see, all of the activities that we associate with behavior in higher organisms, from their perception of the environment all the way to how they act upon those perceptions, are detected and processed by these organisms' nervous systems. Plants, however, do not have anything like a "traditional" nervous system—or any other system, for that matter, that at first glance would allow them to process whatever environmental signals they perceive. This problem did not deter a cadre of innovative, out-of-the-box, and, yes, brave thinkers, who began talking about the "nervous life of plants." And in time, the scientific recognition that plants display true behavior culminated in the emergence of a field we now call *plant neurobiology*.

There is an interesting series of facts that gives particular traction to the idea of plant versions of neurobiological processes. Briefly, the very word "anesthesia" means "loss of sensation," and anesthetic agents are specifically designed to abolish consciousness, sensations, or both, usually in order to perform medical procedures. Studies demonstrate that anesthetic agents inhibit the activity of a variety of cellular processes, and do so in everything from microorganisms to multicellular entities.[8] This hints at fundamental mechanisms conserved over the course of evolution—perhaps not so odd, as animals and plants share many fundamental proteins that control movement, growth patterns, and so on.[9] What *is* surprising is that many species of plants are notably sensitive to anesthetics. Plants exposed to anesthetic agents display changes in their movement, as might be expected, but they also experience changes in seed germination and in that quintessential plant activity, photosynthesis. There is even some speculation about a possible ecological relevance of anesthetics in nature, as plants under environmental stress synthesize ethylene and ether, two well-established anesthetic agents.[10] Can you imagine having the ability to wish away a paper cut, or the pain of childbirth? Remember, many of the discoveries related to anesthetic agents began with compounds isolated from plants. In animals, anesthetic agents work by disrupting the bioelectrical activity of nerve cells, rendering them unable to transmit painful stimuli. Plants display bioelectrical phenomena reminiscent of animal nervous system signaling, and amazingly, anesthetic agents disrupt these plant signals as well.[11] (Still, "plant neurobiology" remains an area of considerable controversy, which I address at more length at the note cited here.[12])

The unarguable reality is that plant behavior exists, regardless of how we decide to define it or the specific mechanisms they use to manifest these phenomena. Plants without a doubt react to their environment, and they also seem to be capable of engaging in processes very much like memory and learning. These are uncontested facts that virtually everyone in the scientific community accepts; I know of no exceptions. There is less of a consensus about whether to describe this behavior as being a consequence of intelligence or even intention, at least as we traditionally define these concepts. I, for one, acknowledge the need to use anthropomorphic language when describing plant behavior. We have no other choice, at least until we expand, redefine, or perhaps invent entirely new terms to describe these activities. On the other hand, I assert that plants deserve better than being seen solely through the lens of animal life, and I am hardly the only scientist who thinks so.[13]

Even our friend Dr. Lindsay expressed frustration that the lack of precise definitions hampered what he (and others) considered the proper treatment of the putative mental properties of plants. In *Mind in the Lower Animals*, in a section titled "Faults of Terminology," he wrote:

> *The faulty or unsatisfactory character of current definitions of metaphysical terms is freely admitted by metaphysicians themselves. The extreme difficulties of the definition or application of the terms used in modern mental philosophy have been pointed out by authors differing so much in their various points of view as Darwin, Lewes, Laycock, and Bain. Lewes, for instance, refers to the "deplorable and inevitable ambiguity of communication resulting from an absence of strictly defined technical terms" as constituting one of the many difficulties which lie in the way of psychological investigation.*

I do agree with another scholar, Dr. Daniel Chamovitz, when he states that "intelligence is a loaded term."[14] He argues that in place of plant "intelligence" we should talk about plant "awareness," awareness being a capacity plants unambiguously have, given that there is no question that they are able to perceive their environment and to act upon those perceptions. At any rate, regardless of the number *or* names of the senses, behaviors, and abilities that we members of the animal kingdom agree to "grant" to plants, the reality is that they use them all—and in all likelihood many that we do not know about—to expertly navigate the intricacies of life on this planet. Plants are master survivors.[15] And nowhere is their tactical mastery more evident than in the world of chemistry.

## OF APPARENTLY USELESS COMPOUNDS

To say that plants are efficient at manufacturing chemicals is a vast understatement. Every plant species displays a degree of chemical "expertise" that is literally billions of years ahead of what the most knowledgeable human chemist can muster. Virtually every substance that plants make consists of small organic molecules; they use every conceivable trick of organic chemistry to manufacture an unfathomably varied number of them. These tricks of organic chemistry are largely still unknown to us—while we have learned to synthesize most of these naturally occurring products, we often are able to do so only very inefficiently compared to plants, with nothing like the finesse displayed by plant biochemistry. Perhaps we should not feel so bad about this; after all, our knowledge of systematic organic chemistry is only two hundred or so years old.

Paradoxically, many of these plant- and fungi-produced compounds do not seem to serve any purpose to the organism that makes them. This flies in the face of what we know about biological life, as organisms survive in no small part by carefully allocating their precious captured energy in order to run their physiological processes. In nature, it is rare to have energetic resources to spare, and simply stated, chemicals are metabolically expensive. It stands to—evolutionary—reason that if an organism invests its precious resources into synthesizing a substance, that chemical must be useful for its own biological functioning. It makes sense for a plant to invest energy to make photosynthetic pigments (like the several varieties of chlorophyll) as these compounds are the reason plants are able to capture light and its associated energy to power up their cellular activities. Yet despite the biochemical cost, many plant species use a significant fraction of their hard-won energy to manufacture chemicals they do *not* use as part of their own life processes. Plants routinely make substances comparable to chlorophyll in both molecular size and complexity without any self-evident usefulness—molecules like nicotine, caffeine, and cocaine, for example, do not play any obvious role in their "mother" plants.

The truth is that these seemingly "useless" compounds made by plants and their ilk in fact have a very specific purpose—namely, what I like to call "warfare, plant style." Some of the most inventive and efficient defensive strategies plants use capitalize on their aforementioned deep mastery of chemistry to manufacture nothing less than highly effective chemical weapons! And the diverse nature of these defensive chemicals is nothing short of outstanding. The fact that some at first don't seem like weapons at all makes them all the more effective. Plants produce substances that, depending on the target, can affect a predator, like a grazing animal, in a variety of ways. There are substances that attract pollinators (a crafty passive weapon that enhances the survival of the plant yet does not harm its target)* as well as substances that might induce irritation that, although not fatal, is usually enough to send the predator on its way posthaste. Some plant-produced chemicals induce changes—profound or subtle—in the behavior of the target. And some, yes, are not just lethally efficient but efficiently lethal: such substances more often than not cause an agonizing death, which may or may not be mercifully swift.

The idea of plant chemical defenses has been "mainstream science" since at least the late 1800s. In the words of the German botanist Christian Ernst Stahl:

> Nobody doubts that plants developed morphological features towards animals during their struggle for life. This is especially evident by the great diversity of flower forms. In the same sense one should interpret the multitude of phytochemicals, namely, that the animal kingdom affected not only the morphology but also the chemistry of the plant.[16]

Chemical defenses are widespread in nature in general, though historically speaking this aspect of nature has been much better

---

* The case of pollinators is a perfect example of how plants cleverly lure animals in order to get them to do their bidding. Plants attract bees, butterflies, and even hummingbirds, who feed on the nectar that the plant produces, and in the process, get coated with the plant's pollen. When the pollen-coated pollinator goes to feed on another plant, the pollen is transferred to that other plant, likely fertilizing it, and the circle of life continues. Some of the substances that plants use to attract pollinators are the very-well-known psychoactive compounds caffeine and cocaine; we'll talk about this in various places in this book.

known in animals than in plants. We know of many (and I mean *many*) examples of venomous animals,[17] and many of these—from spiders and centipedes to snakes—we know intimately, as they are common residents of even our modernized human habitats. Several of these organisms are the subjects of phobias that affect large numbers of people. The fear of snakes is a well-known example; it seems this phobia has been part of us since time immemorial as an instinctive behavior in primates.

Plants, on the other hand, do not strike fear into the hearts of most of us—*Plants on a Plane* doesn't sound like a very exciting movie—and it is easy to understand why. Unlike the jellyfish mentioned in the last chapter, plants do not generally possess complex mechanisms to deliver their toxins, and again, plants do not tend to move much. When they do move, they do so slowly, at least compared to most animals. Even carnivorous plants do not chase the animals that they capture for nourishment. Instead, they rely on their mastery of chemistry to get the job done, emitting chemical signals to attract insects, with predictable consequences for our six-legged friends. Interestingly, in an amazing display of an ability we would not expect plants to possess, certain species of carnivorous plants know to spare the insect species that pollinate them.[18]

It is true that, compared to animals, the compounds plants make for defense and other purposes are simpler; no plant has ever been discovered to produce sophisticated, complex protein toxins like those found in animal venoms. However, simple or not, a plant-based toxin can kill you just as dead as something fancier, molecularly speaking, and what the compounds synthesized by plants may lack in complexity, they more than make up for in sheer numbers.

## Of Undreamed-of Numbers

There are about 300,000 plant species described by botanists, and every year we discover about 2,000 more. In this case, I wasn't including fungi—there are approximately 120,000 described species of fungi, and we discover about 1,000 more every year. In fact, certain studies estimate that there are a whopping *5 million* or so potential undiscovered fungal species.[19] Amazingly, in addition to the "standard" chemicals that plants produce (such as chlorophyll), almost every one of these species synthesizes a variety of unique bioactive chemicals, compounds that serve a variety of purposes. The diversity of the plant world makes for a diversity of compounds, as they don't all produce the same ones; plant families tend to specialize (for example, I know of no plant able to make caffeine *and* nicotine).

Based on the number of plant (and fungal) species known to science so far, there is little doubt that the number of discovered and undiscovered plant-derived substances challenges the imagination. But we can do a bit more than merely imagine; we can illustrate just how many possible compounds might conceivably exist by relying on a few comparisons. Our minds did not evolve to easily grasp the size of very large numbers, but comparisons can help us get a sense of scale. For example, when we think of a billion (1,000,000,000, otherwise known as $10^9$—note the nine zeros), it is easy to lose sight of its sheer magnitude. Think of a billion seconds. Now, what if I told you that a billion seconds is equivalent to about 32 years and change? Suddenly you have a better sense of just how many a billion is. Our universe is about 14 billion *years* old. This means that the number of seconds that have passed since the origin of the universe is about 441,569,000,000,000,000 or about $4.4 \times 10^{17}$, otherwise known as about 442 *quadrillion*.

Now suppose that we want to estimate how many different small organic molecules can be made with 30 carbon atoms (nothing special about the number 30; this is an arbitrary choice based on a typical molecule that is not too big, not too small, but just right, like the legendary porridge*). Thankfully, other scientists have done these calculations for

---

* I know that in the story "just right" referred to the porridge's temperature; please bear with me! (See what I did there?)

us.[*][20] The bottom line is that, based on the rules of organic chemistry the plant world is bound by, there are potentially $10^{60}$ (or 1,000,000, 000,000,000,000,000,000,000,000,000,000,000,000,000,000, 000,000,000) different organic molecules that could be conceivably made out of a mere 30 carbon atoms. (And, of course, the more atoms in a molecule, the higher this estimate will be.) Now, this does not mean that plants do or will produce this number of compounds; rather, this is an illustration of the vast number of *potential* compounds—as more plant species are discovered and evolution makes further alterations, who knows where that potential will take us?

## OKAY, BUT WHY?

We are now in a better position to articulate at least some approximate answers to two burning questions.

*Why do many plants produce substances biologically active in many other species of organisms, particularly animals?*

And:

*Why are animals—vertebrates and invertebrates alike—strongly attracted to many of these chemicals?*

Scientists have proposed a plethora of theories (strictly speaking, more like hypotheses) aimed at explaining why plants make so many of these apparently "useless" compounds (by "useless," I mean not used in their own biological processes), and of the many hypotheses

---

* Please note that there are molecules containing many more than thirty atoms; therefore, in reality, the possible number of organic molecules is much bigger, but I hope that I have made my point.

accounting for the production of various natural substances by living organisms, those seen most often in the scientific literature include:

- that these chemicals are waste products
- that they are opportunistic molecules (that is, compounds that evolved for one "purpose" but are now used for another)
- that they are relics (compounds that were used by the plant in its distant past, but that have no current use—or at least none yet discovered)
- that they are produced for "no reason" (again, that we know of—if a molecule remains part of an organism's biochemical toolbox, it is bound to be useful for something)
- that they are the result of chemical coevolution (more on this in a moment)

One of the best summaries of these hypotheses that I have seen comes from the 2009 book *Nature's Chemicals: The Natural Products That Shaped Our World* by Dr. Richard Firn. This is a beautiful book, plain and simple. It is accessible, reader-friendly, and full of useful information and curious tidbits.[21] I will not go into the details of the hypotheses here, but the main idea is that these compounds were conserved during evolution on account of some serendipitous property—or properties—that proved advantageous to the plant (as defense, offense, or otherwise).

The aforementioned explanations were eventually integrated with the notion of coevolution between plants and animals: a chemical that proved useful in dissuading an herbivore might not stay that way as the herbivore evolves, and the plant might in turn evolve yet another, similar chemical defense to get around this. This integrated explanation—that of conservation of these compounds in the context of coevolution—is the one favored among scientists, probably because of the abundant evidence in its favor, and in fact, this will be our assumption (unless I tell you otherwise) as we talk about the psychoactive substances produced by plants.

To sum up, at least part of the reason for the seemingly unfathomable ability of plants to synthesize so many different compounds is that every organism that has ever called Earth home evolved in close association with other organisms; there are no exceptions to this biological rule. Plants, fungi, microorganisms, and animals interact both with organisms of their same class and with organisms of different classes; most of the time, everybody interacts with everybody else all at once, and—especially in the case of plants, which are limited in their range when it comes to things like sound and movement—much of this interaction takes place via chemicals.

It is easy to understand both the reasons and the mechanisms behind an antagonistic relationship between a plant and an herbivore, but a more subtle topic is the war plants wage on other plants. Many plant species engage in fierce competition with one another, and as you might imagine, chemistry plays a central role in these plant vs. plant skirmishes. However, plants were found to influence other plants well before the discovery of the relevant chemicals involved. Upon the invention of agriculture, it soon became apparent that some plant species did not "get along" with others, meaning that sometimes, when you planted plant "X" alongside plant "Y," one or the other (or both) did not grow as well as they would alone. As early as the mid-1800s, some astute observers, like the Swiss botanist Augustin Pyramus de Candolle,* proposed that some plants "poisoned" the soil for other plants.[22] But it was the 1930s before Hans Molísch, a plant physiologist, not only recorded observations indicating that plants could influence the survival of other plants, but gave this phenomenon a proper name: *allelopathy*, which essentially means "harm toward one another." Presciently, Molísch intended for the term to apply to both antagonistic and mutually beneficial relationships, such as a plant influencing

---

* Yet another medical doctor with a green thumb.

another plant in a beneficial way in order to fight a common enemy, and indeed modern usage of the term includes the action of any chemical secreted by a plant (or by a microorganism) to affect another plant, whether as an enemy or a friend.* More recently, the idea of allelopathy has been applied to organisms other than plants, such as corals and cyanobacteria that make extensive use of chemicals to improve their probability of survival in nature, as well as to the chemicals that plants use against insects (though, as in the case of cocaine and nicotine, these chemicals often find their way to other animal targets as well).[23]

Allelopathy's main idea is that plants can (and do) interact positively or negatively with other organisms in their vicinity; these allelopathic interactions can occur between plants of the same or of different species.[24] In general (and unsurprisingly), negative allelopathic interactions are about competition. In turn, competitive allelopathy can take the form of *additive* or *subtractive* interactions.[25] Somewhat confusingly, subtractive allelopathic interactions are those in which plants use chemicals to *acquire* resources at the expense of other plants (hence preventing their competitors from accessing said resources, these resources being physical space, water, nutrients, light, and so on). Additive allelopathic interactions, on the other hand, are when a plant releases (or *adds*) chemicals to its environment, harming or even killing other plants in the process. Of course, by debilitating or eliminating the competition with these additive strategies, a plant may further its subtractive effort, by virtue of acquiring the newly available resources (as I said, it can be a bit confusing, but surely you get the picture). This is how negative allelopathy works, but of course a great many plant–plant interactions are far from antagonistic; as we shall see, there are many examples of mutually beneficial plant relationships as well. As is the case everywhere in the natural world, some individuals (read, "species") get along better than others.

---

* The usage of "allelopathic" to refer to only antagonistic or both antagonistic and cooperative processes currently depends on the preference of individual scholars, and as such, the "correct" meaning is still debated.

## "THIS IS A MESSAGE FROM THE EMERGENCY BROADCAST SYSTEM"

I grew up in Puerto Rico, which as you surely know, is a geographical region "favored" by hurricanes. I am thus quite familiar with the interruption of television and radio broadcasts by an annoying buzzing sound followed by the ominous lines: "This is a message from the Emergency Broadcast System..." Other regions of the United States are used to hearing this message because of tornado alerts, wildfires, and so on.

But what if I told you that plants came up with the idea of an emergency broadcast system millions of years ago, using messages sent not with sound, but with chemicals? It's true: many plant and fungal[26] species use chemicals to alert their neighbors of "clear and present dangers," like an insect nibbling their leaves. One way they do this is by releasing HIPVs (herbivore-induced plant volatiles), also known as VOCs (plant volatile organic compounds),[27] into their environment. Volatiles are molecules that vaporize in the air (hence the name "volatiles"; every smell you can detect is due to a volatile compound), and these may be released from any part of the plant or fungi, even from the roots, which are usually underground. Neighboring plants and fungi detect these compounds, making them aware of the danger and prompting them to deploy additional defenses. This phenomenon is sometimes (kind of) jokingly called "plants crying for help," and it is widespread; there are thousands of HIPVs/VOCs in close to a thousand plant families.[28] Moreover, it bears mentioning that these warnings needn't be limited to other members of a plant's own species; plants and fungi may use volatiles to alert individuals from other species, allowing them to defend themselves from the common enemy (namely, the aforementioned nibbling insect)—a prime example of plants cooperating rather than competing.[29]

It gets better. Certain plants—notably tobacco and corn, among many others—emit volatiles that serve a slightly different purpose, namely to both dissuade nibbling insects from "nibbling" *and* to attract the offending insects' predators.[30] The idea of plants using volatile chemicals to attract their "enemy's enemies" seems to have been originally proposed close to thirty years ago,[31] but surprisingly, a 2012 article stated that "knowledge of the information content of volatile blends [for this purpose] has advanced little since then."[32] In recent years, this plant strategy has inspired research into the possibility of genetically engineering crops to confer the ability to emit volatiles that attract the natural predators of the pests that plague them.[33] And as we will see, the ability of plants to use chemicals not only to defend themselves via direct means, but also to manipulate the behavior of friends and foes, has direct relevance to the topics of addiction and recreational intoxication.

Plants were here before animals, and so it makes sense that their development of chemical defensive strategies initially focused on other plants. However, likely triggered by the phenomenon of predation by animals (especially insects), plants extended these chemical activities to include defensive strategies against their predators. (As mentioned, the term "allelopathy" can these days refer to plant use of chemicals to target more than just other plants.) Collectively, the compounds plants use to carry out their allelopathic behaviors are called *allelochemicals*. Many of these allelochemicals have been found to mimic natural neurotransmitter molecules in other organisms, allowing plants to wreak havoc on a predator's neurobiology (Figure 4.1). It seems that, over time, certain plants refined their strategies not merely to kill or maim their attackers, but also to take advantage of their enemies' neurophysiology by disrupting their nervous systems. This modification to allelopathy is essentially the origin of how plant substances became psychoactive.

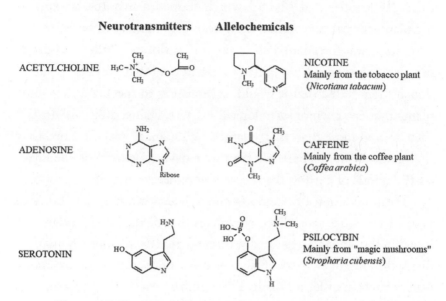

Figure 4.1. Representative animal neurotransmitters and their plant allelochemical counterparts. From Pagán (2005).

## MESSING WITH THE ENEMY'S MIND

The competition for survival between plants and their predators is a natural consequence of the same process that controls the fate of most species in nature, a process defined in various ways over the history of biological thought, but today known to most biologists as *evolutionary arms races*.[34] These coevolutionary races to come out "on top" involve all manner of strategies, chemistry among them, and we have already established that plants are quite skilled at synthesizing chemicals for these purposes. And, in many instances, the compounds that plants make are not about brute force, but finesse. In other words, you do not always have to kill your enemies; sometimes it is enough just to confuse them.

The idea that certain plants are able to mess with the minds of animals appeared in print more than fifty years ago in an article titled "Butterflies and Plants," which framed a hypothesis explaining plant-animal coevolution in light of evolutionary arms races. The hypothesis was formulated using the interaction of plants and butterflies, but the idea can be applied to any animal that preys on plants.[35] The authors stated that it would be "amusing to speculate that some plants practice chemopsychological warfare against animals." In fact they do, and more than that, chemopsychological warfare is precisely the phenomenon responsible for the behavioral effects of chemicals on the minds of entities that possess nervous systems, including us.

The foundation of chemopsychological warfare is the ability of plants to produce specialized chemicals that disrupt the nervous systems of the animals (especially insects) that feed upon them. This disruption might cause confusion or distortions in perception among other effects in order to "distract" the animal away from the plant. An important point about chemopsychological warfare is that it not only works on relatively small organisms like insects, but also on much bigger animals like grazers (think horses, for example), though the effects will obviously differ depending on the animal.*

The idea that plants and fungi are capable of making chemicals to mess with an animal's mind resurfaced in recent work exploring the evolutionary significance of psilocybin, a well-known hallucinogen isolated from a variety of fungi.[36] In this study, the authors stated that "neuroactive compounds like psilocybin that target broadly conserved neurotransmitter receptors may have evolved as a strategy to influence arthropod activity," or as loosely "translated" in the coverage by at least one website, plants "scramble insect brains" and "send them on wild, scary trips."[37] To elaborate with a research-based example, a group of

---

* We will see how this works—in horses, precisely—later on.

investigators from West Virginia University discovered that at least three species of fungi belonging to the *Massospora* genus infect various cicada species. This fact in itself is hardly surprising; fungi infect other organisms all the time. The amazing part of this story is that, having infected the cicada, the fungus goes on to produce (among hundreds of other substances associated with the infection) a couple of well-known psychoactives: the aforementioned psilocybin and a type of amphetamine. *Oh well*, I can hear you thinking, *there goes another confused bug*, but in this case "confused" hardly covers it, and the effects of the alterations in perception are not merely dangerous, but gruesomely fatal. The combination of these two drugs really messes with the cicada's brain in the sense that it becomes (1) uninterested in nourishment (one of the general effects of amphetamines is appetite suppression); and (2) er, shall we say, "obsessed" with the reproductive process. This works out very well indeed for our fungal friend *Massospora* because cicadas infect other cicadas through sexual contact. Alas—and here is the gruesome part—sometimes the drug-addled cicada engages in the sex act so vigorously that its abdomen *breaks off* and remains attached to its partner. The tragically obsessed cicada does not seem to mind; it keeps trying to find things to mate with, despite lacking the anatomical "equipment" traditionally required for such an activity. At least, until it dies.[38]

Whether we will ever know if invertebrates experience drug-induced "trips" just as we humans do is a matter of debate; there is still some (okay, quite a lot of) discussion about whether insects (and, by extension, other invertebrates) possess consciousness.[39] Of course, it is important to reiterate that we are far from putting questions about consciousness to rest when it comes to any animal, humans included. But the specific question at hand here is whether invertebrates in general and insects in particular get "high" on the psychoactive (to humans) compounds that they are exposed to when nibbling on certain plants. Given the close similarity between the invertebrate nervous system and ours, I would say there is a distinct possibility that they do, though much more work is needed before we can approach clarity in this matter. And solving this mystery is not just important for those who are curious about what it's like to be a grasshopper—it is an important part of understanding how psychoactive substances act upon vertebrate nervous systems as well. After all, we are all related to each other through our evolutionary past. We will talk more about the relevance of this fact and its consequences soon enough, but the fact remains that our nervous system and the nervous systems of many of our invertebrate cousins share many structural, physiological, and biochemical similarities. In a nutshell, our sensitivity to many plant substances is essentially a consequence of the basic workings of nervous systems in general.[40]

But for now, even without knowing all the details of how psychoactive compounds affect invertebrates, we can at least say that they do—and have a fair idea of why organisms like the dastardly *Massospora* fungi, above, produce them.

Before we move on to our second burning "why" question—that of why animals, vertebrates and invertebrates alike, are strongly attracted to many of the chemicals produced by plants—let us pause here to summarize what we've learned about why plants produce these chemicals in the first place. In short, it is eminently likely that plants evolved their ability to produce a wide variety of chemicals as a survival strategy. Plants can use chemicals in interactions with other plants, but many of these plant chemicals also act upon the nervous systems of a variety of plant predators, from insects to mammalian grazers. These substances can affect a predator in a variety of ways (good, bad, or very ugly).* The most extreme effect a chemical substance can have on a target is death; an excellent example of this is that many plant-derived substances act as pesticides. In other cases,

---

* One of the "good" plant-animal interactions is pollination. Plants get their pollen widely distributed, while the pollinator gets high-energy nutrients.

the effect of a chemical is simply enough to dissuade the predator from eating the plant, either by inducing unpleasant sensations or by messing with the predator's nervous system in such a way that the predator is no longer a danger (perhaps because it has become drowsy, or ceased to be hungry and wandered off). Remember, there is no single blow that kills a plant, so they can often afford to lose a few leaves before the predator is dispensed with.

Interestingly, sometimes it is difficult to classify a particular substance or chemical as having a specific role. As we saw, the same chemical can kill or even delight depending on the context and the target. An amount of chemical that kills an insect may only cause a mild psychoactive effect in a much bigger human. Moreover, the very same substance can cure or nourish, depending on the circumstances. In the words of Dr. Giorgio Samorini in his 2002 book, *Animals and Psychedelics*, "[T]he boundaries between drugs, medicine, and food are not clear." These ideas and facts about plant-produced compounds relate directly to the further fact that many types of animals seek out these plant-derived substances, and will help to shed light on a variety of the reasons they do so—not least of which is to experience psychoactive effects.

## FOOD, DRUG, OR TOXIN?[41]

It is not surprising that, for a long time, all efforts to understand why organisms actively seek psychoactive substances focused on gaining understanding of the *human* use of such substances. After all, most scientific efforts have human welfare (as well as economic factors) as an implicit fundamental objective. And one of the earliest explanations for the human compulsion to seek and consume such substances was framed in terms of a paradox.

The idea that at the time guided the general direction of most research on plant-human interactions was the *Evolutionary Fitness Consequence Model* (EFCM). The EFCM is based on two basic, complementary assumptions—namely, that pleasurable sensations are coupled with adaptive (survival-enhancing) behaviors, and that nonadaptive behaviors are in turn associated with non-pleasant sensations. For example, contrast the sweetness of fruits rich in vitamin C, an essential nutrient for humans, with the bitter flavor of a poisonous vegetable.* Survival is favored in both cases: the general idea is that the objective experience of adaptive and nonadaptive activities is associated with pleasant or not pleasant sensations, respectively.

This perfectly reasonable, even beautiful, idea flies in the face of a very inconvenient but incontestable fact: many psychoactive plants affect perception, and their ingestion can induce mental states in which there is a "mismatch" between what is perceived and, well, objective reality. Sometimes this "mismatch" is innocuous enough, but any significant decrease in an organism's awareness of reality is an obviously risky proposition (think about a creature that is so intoxicated that it cannot see a big hole in the ground). After all, it stands to reason that in order to successfully navigate the dangers of life, a reliable perception of the environment is essential. Yet psychoactive substances oftentimes also induce very pleasurable sensations. In other words, while the EFCM says that pleasant sensations should be associated with behaviors that are adaptive (that is, confer some benefit), not only do many drugs produce these sensations in the absence of any evident benefit, but in many cases, these pleasurable sensations can interfere with survival itself. Imagine an early human who is heavily intoxicated and therefore unafraid of the saber-toothed tiger stalking her.† It is not hard to deduce the likely outcome: no survival and therefore (as it goes) no reproduction. It is easy to understand the desire of many types of organisms to engage in practices that will result in

---

* Of course, one could argue (and indeed many have) that we learned to consider sweetness pleasant because it conditioned us to ingest vitamin C–rich fruits, and that we learned as a species not to like bitter substances because those who naturally disliked such substances were the ones who survived.
† "Here, kitty kitty . . ."

pleasurable sensations. But why engage in practices that can prevent you from having reliable information about your environment? This apparent contradiction is sometimes articulated as the *paradox of drug reward.*"[42] The "mismatch model" asserts that the paradox of drug reward occurs because psychoactive compounds "trick" the brain. The altered states of mind that humans experience upon taking psychoactive drugs are almost invariably associated with pleasurable sensations. As a consequence of these pleasure-inducing effects—usually associated with adaptive behaviors like eating or sex—humans seek out these drugs, are in turn "rewarded" again by these sensations, and a cycle ensues. According to the mismatch model, this cycle essentially hijacks mechanisms "meant" to reward behaviors that increase survival.* Importantly, it also suggests that addictive behaviors directly linked with drug reward have very likely been with us since prehistoric times and have been enthusiastically practiced by every known human culture throughout history—more on this very soon.

An important consideration to keep in mind is that this seemingly wasteful practice did not originate in a vacuum. Very little in nature is neutral; most of the time, a behavior that persists in a particular species is either beneficial or apparently harmful (there's the paradox again). In the case of psychoactive substances, it seems that the apparent drive to ingest these substances recreationally may have some origin in a drive even stronger than the pursuit of intoxication or pure pleasure, a powerful urge that is of critical importance to an organism's survival. This urge is called hunger.

## The Drunken Monkey Hypothesis

In the late 1990s, Dr. Robert Dudley, professor and chair of Integrative Biology at the University of California, Berkeley, made an interesting

---

* The reward system is a series of nerve cell connections that elicit feelings of pleasure when activated. Interestingly, even though it was initially described only in vertebrate organisms, it has become increasingly clear that invertebrates possess at least something very much like it.

intellectual connection that led to an influential hypothesis about the relationship of our ancestors to alcohol, and the possible evolution of related addictive behaviors. Dr. Dudley, although an established and skillful scientist in his area of expertise, was not a likely candidate to formulate a pharmacological hypothesis, as he was not a biochemist or a pharmacologist. Dr. Dudley studied evolution—more specifically, the evolution of animal flight in terms of biomechanics and bioenergetics. This plays a central role in his story, because the nature of his research oftentimes required him to be in the field, on treetops. During his time "up there," he, along with his colleagues, frequently observed monkeys eating fermented fruit, which was not surprising. However, what *was* surprising was *how* the monkeys were eating the fruit: in an enthusiastic, even frantic way.[43] Moreover, in more than one instance, Dudley and his team observed monkeys eating this fruit so fast and carelessly that they lost chunks of it, dropping them in their feeding frenzy. This flew in the face of what one would expect from a foraging animal; in nature, valuable nutritional resources are rarely wasted. These field observations got Dr. Dudley thinking about alcohol consumption in primates; these thoughts soon led him to certain personal, painful memories of how his family suffered because of his father's alcoholism.[44] Dr. Dudley formalized these intellectual connections in the drunken monkey hypothesis (mentioned in chapter one), which states that "a strong attraction to the smell and taste of alcohol conferred a selective advantage on our primate ancestors by helping them locate nutritious fruit at the peak of ripeness."[45]

The drunken monkey hypothesis proposes that our human affinity for psychoactive substances predates our human lineage, coming to us from our close biological relatives, and that eventually, the consumption of such substances played an important role in our evolution, particularly the evolution of drug-seeking and addictive behaviors. The hypothesis rests in part on evidence that our—really—early ancestors (meaning the common ancestor of chimpanzees, gorillas, bonobos, and ourselves, a species that lived about ten million years ago) already had a significant ability to metabolize alcohol.[46] This fact makes sense only if very early in its history members of this evolutionary line crossed paths with ethanol with some frequency in their daily lives. Otherwise, the ability to efficiently metabolize alcohol would likely have been "deleted" from these genomes by evolution millions of years ago.

We saw in chapter one that the relationship between many types of animals and ethanol, usually from fermenting plant material, is ancient, and it is likely that the habit of seeking and eating fermented fruit did indeed play a role in the survival of our own evolutionary lineage. While I doubt this was at the top of your mind the last time you enjoyed a glass of champagne, alcohol is a very efficient nutrient. It has a higher caloric content than carbohydrates—and in nature, any source of abundant calories is understandably coveted. And most fruits (common providers of alcohol, especially in the wild) are excellent sources of vitamin C, a nutrient that we humans cannot make by ourselves and must obtain elsewhere, usually from food.[47] Alcohol also stimulates certain taste receptors, particularly in mammals, producing pleasurable sensations that without a doubt contributed to the development of our fondness for fermentation. These combined facts turned out to be a very good thing for us. In essence, it seems likely that smell, flavor, and psychoactive effects coalesced in order for us to evolve a liking for fermented food and drink.*

An important detail about the drunken monkey hypothesis is that it implies that alcoholism is a negative consequence of an adaptive behavior (something good for the organism) gone awry. This is similar to current ideas about dietetic habits—many such habits that are

---

* A related aspect of this relationship, of course, concerns the question "What's in it for the plant?" One possibility is the dispersal of seeds by the animal, as in the animal eating the fruit and then depositing the seeds far from the plant, through the usual way, among other factors. Then there is the matter of pollinators; it stands to reason that alcohol might be part of the arsenal of strategies that plants use to attract animals to do their reproductive bidding.

harmful today were beneficial to our ancestors because food resources were not as plentiful then. Finding sustenance in preagricultural and preindustrial times was no trivial matter; it made sense for our ancestors to consume fatty or sugary foods whenever available, as their caloric content might well be the key to survival. And as the availability of such foods was a relatively rare occurrence, their metabolism could take it. In contrast, many (though, sadly, by no means all) people in wealthy modern societies do not lack for food—quite the contrary—meaning that today many of us consume calorie-rich nutrients on a regular basis while metabolizing them using pathways not evolved for times of plenty. As a result, diseases that are heavily influenced by the body's biochemical handling of nutrients—including obesity, heart disease, and diabetes, among others—are now present in almost epidemic proportions in many modern cultures. In the same way, an excessive (and sometimes obsessive) affinity for ethanol might have been advantageous to our ancestors thousands of years ago, but as it developed over time into what we know as alcoholism, this proclivity clearly became harmful.

The drunken monkey hypothesis, although logical and even attractive, raises more questions than answers. For one thing, the issue of nonadaptive altered states rears its head again, as it is easy to imagine that hypothetical drunken monkey falling off a cliff (or disregarding evident danger—like the meme monkey from chapter one, about to whack a lion with a stick). What's more, alcohol is still a toxin, and not only is too much bad for you, *way* too much will kill you. Dr. Dudley and his collaborators acknowledge the questions raised by their hypothesis (most good hypotheses *do* raise further questions),[48] and other scientists are putting this idea under close scrutiny.[49] What is indisputable is that the drunken monkey hypothesis is an important contribution to our quest to understand the phenomena of alcoholism in particular, and addiction and drug-seeking in general.

## The First Nutraceuticals*

In ancient times, we ate for a simple reason: because we were hungry (not just peckish, either; most of the time we were *really* hungry). And we were not very discriminating about what we ate—we couldn't afford to be; survival was at stake! What we did not know at the time was that, in addition to the much-needed calories that we got from food, we also acquired nutrients (for example, vitamins) that are essential to the proper function of the body and yet that we generally cannot make ourselves. In modern times, thanks to our relatively sophisticated knowledge of biochemistry, we are aware both of the existence of these nutrients and where to get them. We make a special point of doing so, too—thanks to the unfortunate populations before us who suffered from diseases (like scurvy or rickets) eventually traced to lack of certain nutrients, we are aware that their deficiency will invariably result in some kind of malady.

Sometimes, though, the nutritional benefits of a substance are not immediately apparent. This is well illustrated by psychoactive compounds—one does not usually think of cocaine, nicotine, alcohol, or other abused drugs as food. Of course, having just heard the details of the drunken monkey hypothesis, it will not surprise you to hear that some scholars argue convincingly that alcohol was a significant source of calories for ancient peoples (and likely for other primates as well).[50] But while in the case of alcohol it is easy to see the apparent advantage of its consumption as a direct source of chemical energy, the connection between other commonly abused substances and nutrition is not as direct.

The mismatch model and its accounting of the paradox of drug reward is of course not the only idea scientists have come up with to explain the ingestion of psychoactive substances by humans and other animals. Some authors propose an explanation that blurs the

---

* A "nutraceutical," in case you are wondering, is any food element used for medicinal purposes.

lines between medication and nutrient.[51] This alternate model, which we'll call *the nutritional hypothesis*, takes into account a variety of facts and hypotheses. For example, there are societies whose consumption of plants includes tobacco and coca (the main sources of nicotine and cocaine, respectively) for nutritional purposes and not necessarily because of any apparent psychoactive effect (we all have to eat, after all, and hunger is a universal condiment). We've already talked about the fact that many psychoactive substances are allelochemicals, used by many plant and fungal species to engage in a so-called chemopsychological war against predators, and that many of these allelochemicals closely resemble natural neurotransmitters in animals. Well, as it happens, the relationship between some psychoactive substances and nutritional needs seems to be a direct consequence of their allelochemical properties.

## ALKALOIDS: SECONDARY, BUT FAR FROM UNIMPORTANT

Most psychoactive drugs, and a significant fraction of allelochemicals in general, are classified as *alkaloids*. A detailed explanation of what a compound has to look like to be classified as an alkaloid is beyond the scope of this book, and at any rate, even among organic chemists there are sometimes disagreements about whether a particular candidate fits the bill. For our purposes, it is enough to grasp a few fundamental characteristics—the most fundamental being a molecular makeup that involves the presence of a nitrogen atom as part of a ringlike structure (as mentioned earlier, nearly all of the familiar "toxines" we cover in this book are alkaloids). Alkaloids also tend to be basic (alkaline) as opposed to acidic—hence, "alkaloids." One of their more general characteristics is a bitter taste, which is likely part of the reason plants make so many alkaloids in the first place, so that the taste might deter predators from grazing upon them.

The number of currently known alkaloids is somewhere in the order of twenty thousand or so,[52] and chemists discover (and synthesize) new alkaloids frequently, so by the time you read these lines it is likely that there will be more than there were when I wrote them. Most known alkaloids come from plants, fungi, or microorganisms, and in virtually every case, these compounds form a central part of their defensive or offensive survival strategies. Alkaloids also tend to be the products of *secondary metabolism*, a term coined in 1891 by the Nobelist Albrecht Kossel. The resultant compounds are oftentimes called *secondary metabolites*, and in this context, "secondary" means that they are specialized as opposed to general-use compounds—for example, not all plants make caffeine, but all plants make the compounds ATP, chlorophyll, glucose, and so on. These metabolites do not have an apparent function in the biochemistry or physiology of their parent organism, which brings us to a key point: there are compounds that meet all the structural "requirements" to be considered alkaloids, and yet are not called alkaloids. These include neurotransmitters and hormones, which of course *do* play important roles in the biochemistry of the organism that produces them, and this is an important distinction between, say, nicotine and a common neurotransmitter like serotonin. In summary, we can say that an alkaloid (generally, and for this book's purposes) is a compound having the aforementioned structural features, but which does *not* play a direct role in producing an organism's biochemistry or physiology.[53]

Let's look at how this works by imagining a hypothetical insect. When this insect nibbles on a tobacco plant, it is bound to ingest a certain amount of nicotine. There are two primary possible outcomes of this ingestion for our insect: the first of these is that it might get "confused" or "high" (admittedly anthropocentric terms),* causing it to wander or fly aimlessly away from the plant. Alternatively, if the amount of nicotine it ingests is large enough, the insect will die (the original "purpose" of nicotine being to serve as a pesticide).[54] The reason nicotine disrupts an insect's physiology is because it interacts with specific targets, most of them receptors that form an integral part of nervous systems and control the all-important process of neurotransmission.[55] In essence, a plant-produced allelochemical (nicotine) mimics the structure of a natural neurotransmitter (in this case, arguably the best-known one: acetylcholine[56]) within an insect's brain in order to inactivate, or in some cases "hyperactivate," a particular receptor system.† Either way (assuming sufficient dosage to get the job done), this interaction has dire consequences for the insect.

Now, instead of our hypothetical insect, let us imagine an ancient human who happens to be rather hungry—along with some degree of desperation, he is likely also suffering from nutritional deprivation. It is no surprise that malnutrition causes a decrease in many important biomolecules; neurotransmitters, acetylcholine included, are metabolically expensive. Thus, in our hypothetical malnourished human, acetylcholine (which is integral to the function of the nervous system in a number of ways) would likely be depleted.[57] If this poor guy, desperate for sustenance, munched on a tobacco plant, some nicotine would have made its way into his system, and some of this nicotine

---

* This is the very working definition of a psychoactive substance. "Psychoactive" literally means "mind-altering."

† In general, nicotine may either inactivate or "hyperactivate" the nervous system of an insect. In vertebrates, including humans, nicotine is invariably a "hyperactivator" of the nervous system.

would have reached the "hungry" acetylcholine receptors in his brain and at least partially activated them; the now-active receptors would be able to function properly once more. The happy outcome of this story is that when our ancient hero ate those tobacco leaves, not only did they alleviate (at least a bit) his hunger, but the activation of his acetylcholine receptors by the nicotine he ingested in the process *made him feel better*. From a moment like this one, according to the nutritional hypothesis, people learned that some plants were more than food; some would also make them feel better (and, as we all know, sometimes a bit loopy as well—a trade-off or perk, depending upon perspective).

This story could well have been told about cocaine, morphine, cannabinoids, or even hallucinogens, among many other substances. All of these compounds mimic other neurotransmitters that are also essential to the function of our nervous system. And despite the "origin story" of most allelochemicals (their effect as pesticides), the amount that people would have consumed when eating the plants that produce them would not generally be enough to kill a human—after all, even an underweight human weighs hundreds of thousands times more than your typical insect. Amazingly, we can often refer to the exact same substance as a nutrient, a medication, a toxin or a poison, and a recreational molecule. It all depends on the compound, the context, the intended or unintended "victim," and so on.

Over the course of this book, we've looked at any number of explanations for how the relationship between animals (including humans) and drugs began, and now for *why* it began (and persisted). It is important to remember that both the nutritional hypothesis as well as the mismatch hypothesis are, well, hypotheses, both with ardent proponents and equally passionate detractors; though both are based on facts, they are only possible explanations for those facts, and neither should be taken as revealed truth (no hypothesis should). Moreover, these two hypotheses are not mutually exclusive, as both might well be correct to a certain extent—there might also be factors we have not considered that warrant the formulation of entirely new hypotheses. At the end of the day, there is still a lot we don't know about the ancient and ongoing interactions between animals and plants.

---

No organism that exists can do so independently of other organisms; every single living thing associates with other living things. This is one of the most evident characteristics of biological life. The nature of these relationships varies; predators and prey, cooperators and partners, sexual mates and sexual competitors, thieves and victims . . . the list goes on. Sometimes these relationships are antagonistic, sometimes collaborative, and sometimes, as with human relationships, "it's complicated."

There is no question that plants are living organisms, yet we sometimes lose sight of this fact—the very term "vegetable" implies inactivity and unresponsiveness, but as we've seen, neither of these are inherent or even usual characteristics of plant life. Plants display ingenious adaptations that, besides being interesting to study, can lead to a host of applications. In other words, the many survival strategies used by plants and fungi may well help to ensure our own welfare and survival.

But the obvious advantages we've gained by making use of various plant-based chemicals is not the only reason plants are worth our attention. In our evolutionary story in general, and the specific story of psychoactive drug use by animals, the active role of plants and fungi is oftentimes overlooked. Yet we can say with a significant degree of certainty that plants and fungi rule most aspects of our world. And not one of the stories in the two chapters that follow would be possible without their ruthless (and very successful) drive to persist.

# CHAPTER 5

# SPINELESS MINDS ON DRUGS

*Like taxes, the fly is always with us.*

—DR. VINCENT GASTÓN DETHIER, *TO KNOW A FLY*

*The worms will provide the fascination—it is
up to you to provide the science.*

—DR. JAMES V. MCCONNELL, *A MANUAL OF
PSYCHOLOGICAL EXPERIMENTATION ON PLANARIANS*

It seems self-evident that it is not wise to stir up a hornet's nest. There are even common sayings using this as a metaphor, with the general idea being that if you engage in such a foolish practice, you will not like what comes next. This is true not just of metaphorical hornets' nests but in reality, as many hornet species are quite aggressive. That said, in 1969, some brave scientists took it upon themselves to stir up some members of a species of social hornet, *Vespa orientalis*, by dosing them with a variety of drugs, including (among others[1]) LSD, antidepressants, tranquilizers, and, of course, ethanol.[2] Given their usual temperament, you might expect hornets to be angry drunks, but the results of this experiment did not show any alcohol-induced increase in aggressive behaviors. In fact, the authors of this study described quite "subdued" behaviors in their subjects across the board. This was most likely because they needed to use relatively high concentrations of the tested substances in order to observe any effects at all—a hunch supported by the fact that each of the substances they tested has the potential of causing drowsiness at high concentrations. To date, this is the only such study I could find, and so the scientific world awaits the day when another group of fearless souls feels called upon to see what an intoxicated hornet gets up to. For now, perhaps we should amend the popular saying to note that if you *are* going to stir up a hornet's nest, you ought to get the hornets good and drunk first.

As we'll see shortly, things went rather differently when a group of scientists gave alcohol to members of another undeniably aggressive insect species. Of course, the very definition of the concept of aggression is a contentious issue (most behavioral definitions are); after all, we cannot possibly know if a group of insects is actually angry or if there are other factors that make them more prone to attack and sting—for instance, an increased "sense" of territoriality. Nonetheless, these studies offer us some information about how psychoactive substances affect behavior in our invertebrate cousins. And, as we'll see in dramatic fashion many times, the similarities between us can be both surprising and illuminating. But understanding invertebrates better has value beyond how it might illuminate the lives of humans; just like plants, these creatures are worthy of consideration in their own right. The stories in this and the next chapter should be of interest to anyone curious about the experiences of the other animals with whom we share our planet.

## DRUNKEN KILLER BEES, AND OTHER TALES FROM THE HIVE

In the 1950s, scientists from Brazil were studying ways to boost honey production. They came up with an approach that involved creating hybrids of European and African honeybee varieties, since, while the former were those most commonly used by beekeepers (on account of their being less aggressive), the latter generally produced more honey. Sure enough, the hybrids inherited the African variety's increased honey-producing capacity, but unfortunately, they also inherited its short fuse and general nastiness. These hybrids are known as Africanized honeybees, and they are *not* known for their gentle demeanor—in fact, you may have heard them referred to by a nickname: *killer bees*. One fine day, in a real event that yet would not have been out of place in a science fiction movie, several hybrid colonies broke through quarantine, escaped, and began migrating north while "going forth and multiplying." The first members of these rogue colonies were detected in the United States in 1985, and there is every indication that these bees are here to stay. While the massive bee-pocalypse predicted by the media when news of the escapees (escap-bees?) first broke has (unsurprisingly) not come to pass, these bees are quite aggressive and they have indeed killed people, not because their venom is especially toxic, but because of the viciousness of their attacks: they go after a perceived threat en masse, and continue attacking for longer than "ordinary" bees.[3]

Dr. Charles Abramson is a comparative psychologist at Oklahoma State University who has studied the effect of ethanol on bees with the objective of using them as an animal model for alcoholism. Along with an Oklahoma collaborator, Dr. Aaron J. Place, and Drs. Italo S. Aquino and Andrea Fernandez from the Universidade Federal la Paraíba in Bananeiras, Brazil, he set out to see what happens to Africanized bees on ethanol.[4] The researchers' rationale was that the link between alcohol consumption and aggressive behavior in humans is well established (scientifically and colloquially); it was hoped that using honeybees as an animal model might shed light on various aspects of alcoholism, including aggression. They thus decided to measure a behavior that might be linked to aggression in bees, namely the extension of its stinger. They chose to work on killer bees in part because of previous results showing that European bees given ethanol did not display any increase in their rate of "stinger extension."[5] Dr. Abramson and his gang reasoned that Africanized bees, by virtue of being grumpier, would show aggression in this manner more frequently.

The first iteration of their experiment involved harnessing individual bees, fixing them in place, and administering the relevant experimental solutions (i.e., spiked or non-spiked) via a small needle. They then counted the number of times the bees "showed their weapon" over a subsequent period. Alas, Africanized bees did not behave any differently than their tamer cousins had, at any concentration of ethanol. For example, out of 1,500 independent observations, 1,468 bees (about 98 percent) did not show stinger extension. After some brainstorming, the group hypothesized that the harnessing technique they were using might physically interfere with the ability of a bee to move its body in the way it must to extend its stinger; forcing the bees to "drink" via injection could potentially be affecting their behavior as well. Thus, they redesigned their experiment in two key ways: they used free-flying bees, and they allowed them to drink whenever they wanted by placing sucrose solutions with or without 20 percent ethanol within reach of the colonies. After giving them access to the solution for about an hour, the scientists proceeded to taunt the bees by dangling a piece of leather in front of the hive. The bees obliged by stinging; scratch that, they attacked those leather patches with relish. The idea was to count the stingers left in the leather patches as a way to measure sting frequency.*

The experiment's design called for about seven hours of observation. Seven hours is a long time to look at a bee, and nothing of consequence happened for the first four hours, so they must have been feeling rather disappointed. That all changed during the fifth

---

* When bees sting, their stingers stay stuck in the target; when the bee tries to pull out, part of its insides are usually left behind, most notably the venom gland, which keeps pumping the goods into the (unlucky) target (not that the bee is so very lucky in this scenario, either).

hour, when, according to Dr. Abramson, "bees that fed on an ethanol solution acted very aggressively." In fact . . . well, let's let him tell it, shall we?

*The bees began to sting so vigorously that the experimenter dangling the patch became alarmed, and bees breached the laboratory and began stinging people inside the building.*

I know. It sounds biblical.

Incorporating the lessons learned from that experiment, they conducted additional experiments using a time period of five hours (and presumably beefed-up lab security). The results indeed showed that a solution of 20 percent ethanol seemed to increase aggressive behavior in killer bees, as measured by the average number of stings in the leather targets (nine and seventeen in the absence and presence of ethanol, respectively—a statistically significant difference). These researchers were undeniably a committed bunch; it takes dedication to purposefully anger a hive of killer bees for the sake of scientific discovery.

## STUNG IN THE NAME OF SCIENCE

Perhaps some bravery is a requirement for work with stinging bugs. Dr. Justin O. Schmidt, an entomologist at the Southwest Biological Institute of the University of Arizona, is undoubtedly a brave man, as evidenced by one of his claims to fame: the well-known *Stinging Insects Pain Scale*, which is exactly what it sounds like.[6] This scale rates insect stings according to how much pain they inflict on humans—well, on one human, anyway. You see, Dr. Schmidt's invention of this scale was no mere theoretical exercise; he created the pain scale from direct personal experience by allowing (and sometimes actively encouraging) a wide variety of insects to sting him. The scale goes from 1 to 4, with 1 meaning "barely felt," 2 "hot," as in the sting of a cigarette burn, 3 feeling as if "after eight unrelenting hours of drilling into that ingrown toenail, you find the drill wedged into the toe," and 4 . . . well, according to the brave doctor, a level 4 sting is "pure, intense, brilliant pain. Like walking over flaming charcoal with a 3-inch nail in your heel."[7]

The current pain champion among insects is the appropriately nicknamed bullet ant (*Paraponera clavata*), which hails from pretty much everywhere in South America and was discovered by science back in 1775. This peaceful ant (it does not attack unless provoked) rates a full, definitive, and unapologetic 4 on the pain scale. In fact, its sting is so painful that the Mawé, a South American tribe known to treasure braveness, uses this insect in a rite of passage. At about age twelve (!), Mawé boys endure the pain of bullet ants affixed to woven gloves that the preteen must wear for a certain period of time—this is somewhere in the range of ten to thirty minutes, depending on the source. This "procedure" must then be repeated between twenty and twenty-five times, after which the boys are considered men.[8]

As of 2018, Dr. Schmidt has been stung by some 150 different species of insects (talk about job dedication!). I have read some articles that characterize his efforts (and him personally) as "crazy"—I, for one, do not share that opinion; he is a learned scientist, who happens to be a braver person than most. That being said, I would not engage in this type of research myself, nor encourage my readers to do so, in no small part because you never know what you will turn out to be allergic to.[9] However, I would encourage you to read his very enjoyable book *The Sting of the Wild*, in which he narrates many fascinating stories of his favorite stinging critters. The book also tells the story of another brave entomologist, Dr. Michael Smith.[10] Dr. Smith refined Dr. Schmidt's pain scale, also using himself as a research subject, with two important differences: he rated only honeybee-induced stings, and he allowed (encouraged?) said bees to sting him in various . . . unexpected areas of his anatomy (Figure 5.1). To give you a taste of the paper publishing his results, in a prominently displayed disclaimer, Dr. Smith states:

> *Cornell University's Human Research Protection Program does not have a policy regarding researcher self-experimentation, so this research was not subject to review from their offices. The methods do not conflict with the Helsinki Declaration of 1975, revised in 1983. The author was the only person stung, was aware of all associated risks therein, gave his consent, and is aware that these results will be made public.*[11]

I will leave you to discover more about this other dedicated soul for yourself. You will be either utterly fascinated or totally freaked out (both?) by his tale, but it is well worth the read either way.

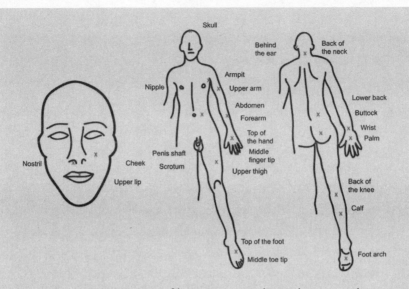

Figure 5.1. Diagram of bee stings in order to determine the perceived pain by location. From Smith (2014), reprinted by permission of the author and under a CC BY 4.0 license (Yes, the author of this paper did this to himself, on purpose).

## Bouncer Bees?

We have explored the fact that the natural phenomenon of chronic alcohol intake (or that of other psychoactive substances) is a paradoxical behavior in the sense that it often impairs the ability of that animal to function. This paradox is even more pronounced when it occurs in the context of a bee colony. Bee colonies, you'll recall, are superorganisms—super-organized societies that function collectively, with a division of labor and regimented behaviors that almost never deviate from established norms. The worker caste of social insects such as bees is tasked with a crucial job: to actively seek food for the sustenance of the hive. Scouts fly considerable distances trying to find suitable food, and when they return they communicate the location, type, and quality of the food source by performing various behaviors, some of which include elaborate dances.[12] Sometimes, the food source consists of nectar that is already undergoing fermentation and therefore is laced with some concentration of ethanol. As you might imagine, then, it is not unheard of for forager bees to encounter fermented nectar in their travels. But when a bee returns to the hive a bit "tipsy," well . . . let's just say there is reason to believe that the hive "administration" has a zero-tolerance policy for this kind of behavior.

There are many anecdotal reports telling of bees that fly erratically back to the hive and behave for all the world as if they are drunk. A curious detail here is that, in experiments, researchers have found that bees can tolerate amounts of ethanol that, if scaled by weight, would render most other animals unable to function (more about this shortly). Since bees are evidently alcohol-tolerant, and yet they sometimes return to the hive in an intoxicated state after consuming fermented nectar, perhaps there is something *else* in the nectar that affects them? This point is certainly worth researching. Anyway, speaking of research, in experiments where researchers purposely intoxicate their bee subjects, if a "tipsy" bee engages in trophallaxis—the common, if admittedly a little gross, practice of regurgitating food to feed another insect—the receiving bee tends to get "tipsy" as well. You can see how, if this also occurs in nature, it would be bad news for the colony, as it is a bit like the spread of a contagion. You simply cannot run a bee society in an efficient manner if a significant proportion of the bee population is in an intoxicated state. And as bee societies are capable of remarkably intelligent behaviors in furtherance of community goals (recall the public health example we saw at the end of chapter three), it would not be surprising to find them engaging in behaviors that prevent intoxicated bees from "infecting" the hive.

The facts and ideas above are the probable explanation for a semi-mythical story about bees who become so drunk while collecting fermented nectar that when they return to the hive, their nest mates—specifically those that guard the beehive's entrance—reject them, sometimes violently. These bouncers are even sometimes purported to "execute" a drunken bee (perhaps it becomes bee-lligerent?) by dismembering it.*

It seems that this idea can be traced to an article published more than one hundred years ago in the *Manchester Guardian* (now just the *Guardian*).[13] In the piece, the author speculates that bees can get so drunk that they become understandably uncoordinated, making them easy prey for birds. However, there were no reported observations of birds caught in the act of snapping up stumbling bees. This 1917 article was in turn probably the origin of a variation of the tale told in a 2001 interview (also in the *Guardian*) by Dr. Errol Hassan, a distinguished entomologist based at the University of Queensland in Brisbane, Australia.[14]

## A SCHOLAR IN LOVE

This anecdote is off topic, but too adorable to omit. In 2018, Dr. Hassan did something quite romantic for his wife of more than fifty years, Ursula. In honor of her birthday, he named a newly discovered species after her—an especially beautiful type of wasp.[15] About this new species, *Uga ursulahassane*, he said: "It's a beautiful little thing, and it is such a fitting way to pay tribute to someone who deserves to be remembered—for time immemorial." I think this is a very sweet and thoughtful gift, and I wish I could do something similar for my own wife, but I happen to work with planarians, a type of flatworm we'll talk more about soon. This is a matter of self-preservation for

---

* There are videos that purport to show this punishment being enacted, (www .youtube.com/watch?v=ZhUKLsSjUZs), but without the proper "provenance" of the offending bee (knowing whether the bee is drunk or not), this is circumstantial evidence at best, and as justification for this bee-havior (sorry, I will stop now!) it would certainly not hold up in any court of law.

me; I do not think my better half would appreciate having her name associated with a flatworm "for time immemorial."

In the *Guardian* interview, Dr. Hassan stated the following about drunken bees: "They are not allowed in by the guard bees at the entrance to the hive . . . they are pushed away from the landing platform or attacked." Taken at face value, this statement offers nothing to give one pause—a behavior such as the one described would be obviously advantageous for reasons mentioned above. However, I was unable to find any formal scientific report of this behavior, and the fact remains that, in science, one needs the proverbial receipts (which in this case means peer-reviewed publication) or it did not happen. I reached out to Dr. Hassan for guidance and clarification and, while waiting for a response,* I contacted the intrepid Dr. Charles Abramson, whom we met in the story of the sloshed and sting-happy killer bees. Dr. Abramson shared with me that while he is likewise unaware of any formal reports of this behavior, he had of course heard about it as well and even attempted experiments with collaborators in Turkey and Bulgaria in the hopes of solving the mystery of whether this behavior actually occurs. Alas, the results were inconclusive, but *quite* interesting.

Briefly, Abramson and his fellow researchers identified two "sister queens" in separate colonies of an apiary located at Uludag University in Bursa, Turkey. They gave one of the queens (a.k.a. "the experimental queen") a sucrose solution laced with 10 percent ethanol, and gave the second queen (a.k.a. "the control queen") an alcohol-free sucrose solution. The queens avidly drank the sweet offerings, after which the researchers returned them to their respective colonies; both colonies accepted their queen seemingly "without question." There were no measurable differences between the two colonies' day-to-day

---

* None yet, alas!

functioning afterward, except for one detail: the experimental queen was slower to lay eggs. The scientists did not think too much of this at the time; however, two weeks after the start of the experiment, they discovered that the experimental queen was missing—never to be seen again—and had been replaced by another queen! The scientists hypothesized that the bee proletariat had detected the abnormal behavior of their queen and quickly "deposed" her. Oddly, the new queen looked older than her actual age, laid fewer eggs, and eventually the colony itself looked "unhealthy." (Generally, an "unhealthy" colony displays a diminished bee population, depleted honey and pollen stores, and evidence of parasitic infections.) This work provides at least some evidence for the notion that a beehive as a community rejects drunken bees for safety's sake. True enough, two bees hardly a scientific theory make, but hey, you have to start somewhere, right?

## HOW TO GET A QUEEN BEE DRUNK

While honeybees are notoriously (at least among entomologists) resistant to alcohol (in terms of the amount they can consume without dying), there are also studies indicating that they experience changes in behavior when consuming amounts of ethanol that would be comparable to the amount a human might consume.[16] Another reason the queen bee experiment conducted by Abramson and his collaborators is interesting is because it illustrates just how much alcohol bees can tolerate. Dr. Abramson and company observed that the experimental queen drank her 10 percent ethanol-laced sugar solution at a rate of about 2 microliters (µL) per second (a microliter is a millionth of a liter), consuming a total of 20 µL. Now, a typical bee weighs about 0.00025 lbs. If we translate the amount of alcohol the queen drank into human terms, we get the following proportion: 20 µL / 0.00025 lbs = 80,000 µL/lb. For a 200-pound human to consume the equivalent would mean drinking just over four *gallons* (200 lbs x 80,000 µL/lb = 16,000,000 µL or 16 liters) of an alcoholic beverage ... in *ten seconds*. The human stomach can distend up to about one gallon at the most. It is physically impossible for a typical human to drink four gallons of anything in one sitting, let alone to consume four gallons of what amounts to strong beer or light wine in ten seconds. Even accounting for physiological and biochemical differences, this is an impressive amount of alcohol for a bee—queen or otherwise—to consume. The lesson being that just as it is unwise to stir up a hornet's nest, it is undoubtedly inadvisable to play a drinking game with a bee.

## Flying on Uppers

Alcohol is not the only psychoactive substance in the bee pharmacopeia; science has learned a lot from these social insects when it comes to stimulants like caffeine and cocaine.

Cocaine, like nicotine, is produced by plants as an insecticide, and in an attempt to explain why such a substance should activate reward pathways in humans (a conundrum that will be familiar to you from our discussions of drug paradox in the previous chapter), scientists have often resorted to explanations relying on the differences between invertebrates and mammals like us. However, research has shown that bees exposed to cocaine in fact seem to react very much like humans. Cocaine attracts bees both in the wild and in experimental settings, and like humans they exhibit tolerance and withdrawal. In other words, at the right dosage, bees may indeed experience a "reward" effect from cocaine, perhaps because, as is the case in humans, the same chemical pathways are involved in both reward effects and motor control—disruption of the latter system being how cocaine exerts its deadlier effects on insects. That there might exist significant similarities between us and bees in terms of drug response of course makes drugged bees of great interest, particularly when it comes to behavioral effects. Notably, studies have found that cocaine affects the well-known dance behavior that honeybees use to communicate the presence and location of suitable sources of food as well as how well they remember this information.[17]

Another stimulant, caffeine, is the "drink of choice" for many of us to jump-start our day, boost alertness, or prepare for a long session of working or studying. There is a great deal of research indicating that caffeine is indeed effective at enhancing "lower" cognitive functions such as simple alertness or response time when completing a task. Alas, despite caffeine's reputation among laypeople as a general "brain-sharpener," there is a dearth of information—at least in humans—about "higher"-level cognitive benefits, for instance memory enhancement or the improvement of learning processes. But as invertebrate animal models continue luring pharmacologists into their webs with their enticing similarities to us, scientists hoping to untangle the effects of caffeine have turned their attention to nontraditional research subjects. One of these is the European honeybee (*Apis mellifera*).

When honeybees are given caffeine experimentally, a couple of rather curious effects become apparent. Young bees on caffeine learn to associate specific smells with particular tasks faster than similarly aged, caffeine-naïve bees. In older bees, caffeine exposure enhances the ability to learn complex tasks and improves motivation. (I know what you are thinking: *How on earth do you measure the motivation of a bee?* These are the thorny questions science must tackle!)[18] In other words, there does seem to be evidence that caffeine has salutary effects on learning in bees. So far, there is no evidence that caffeine affects a honeybee's memory formation or retention.[19] However, another rather intriguing finding of bee caffeine experiments relates to the slowing down of aging processes. A 2014 paper reported that a series of biochemical markers related to aging were reduced or slowed in bees exposed to caffeine.[20] Don't go rushing an extra cup of coffee just yet, as I have not seen any confirmatory studies; however, the idea of a link between caffeine and aging has made numerous appearances in the scientific literature.[21] And while the jury is still (very much) out when it comes to antiaging effects, the good news is that, even if you are not a bee, caffeinated beverages seem to be good for you overall, when consumed in moderation. As in all things, the dose makes the poison—or panacea.

## THE SEX LIVES OF DRUNKEN *DROSOPHILA*

Ah, you didn't think I'd forgotten our friend the fruit fly, did you? Not possible—after all, what they lack in size, they more than make up for in scientific interest.

In chapter one we were introduced to the inebriometer, created to test *Drosophila melanogaster*'s resistance to alcohol. As useful as this invention proved to be, the only information the inebriometer could give us (at least in a straightforward way) was how a fly flies (or doesn't) upon exposure to ethanol. But what if a curious scientist wanted to observe a more subtle effect related to alcohol consumption, like being unlucky in love? To achieve this goal, adventurous researchers must resort to more sophisticated strategies, since love, as sublime as it is, does tend to complicate things, even in science.

Reproduction is one of the primary "biological imperatives" in nature—in fact, as far as nature is concerned, sex (or another reproductive corollary) is kind of the whole point of living.[22] Thus, it is hardly surprising that mating behaviors are often among an organism's strongest impulses. These behaviors are often reinforced by the reward of pleasurable sensations, while failure to engage in sexual behaviors (because of rejection or other factors) is a well-known trigger of a variety of mostly negative responses by the organism. That being said, it is important to point out that, in many animal species, the sexual act is not pleasurable—or even safe—for one or both of the partners. Perhaps the best known of these dangerous liaisons is a variety common to many spiders as well as insects like the praying mantis. In these species, males bet their own lives in a gamble to pass their genes to the next generation, oftentimes paying the ultimate price when they are summarily beheaded postcoitally. (The things we do for love!) Another dramatic example, in which the male comes out on top, so to speak, is that of the common bedbug. This time, the scientific name says it all: *traumatic insemination*. The male bedbug attacks and stabs (yes, *stabs*) the female with his sexual organ, depositing his genetic contribution directly into her body cavity and bypassing any sexual organ. Only after that violent act does the sperm travel to the female's ovaries.[23] (Unpleasant creatures, bedbugs.)

The reproductive behavior of our fruit fly friends is more "traditional." The male *Drosophila* employs a mating strategy that depends on wooing the female to achieve his objective. Moreover, their behavioral responses suggest that flies experience pleasure (or at least something very similar) in the offing. In order to impress the object of his affections, your typical fruit fly suitor tries various strategies: he may vibrate either wing at particular frequencies as a little serenade, gently touch the female's abdomen, or even (in a behavior that one is tempted to anthropomorphize) nuzzle her genitalia with his "mouth."[24] Similar courtship strategies exist throughout nature's kingdom among vertebrates and invertebrates alike, all because of the following fact: in a significant fraction of sexually reproducing species, it is largely up to the female to decide when and with whom she mates (and rightfully so, I may add).[25] This only seems fair, as in virtually every animal species it is the female who invests the most effort in ensuring the survival and general welfare of any offspring, using precious energy reserves in the process. Thus, the male fruit fly must prove himself worthy; otherwise, nothing of consequence will happen. On average, the male courts the female for approximately ten minutes. If all goes well for our gallant hero, copulation ensues for about twenty minutes or so, hopefully followed by fertilization. After the act, the couple goes their separate ways: the female to lay her eggs, and the male . . . to search for another female. And so the famous circle of life keeps on turning.[26]

Alas, we can't always get what we want, and neither can the fruit fly; some of us are unlucky in the game of love. In 2012, Dr. Galit Shohat-Ophir (then part of the research group of Dr. Ulrike Heberlein) published a report describing what amorous male *Drosophila* do when Lady Luck is not on their side. Dr. Shohat-Ophir and her collaborators divided male flies into two groups, then proceeded to toy with their affections in the name of science. One group of flies was given access and opportunity to mate at will, while the other was prevented from mating. Sometimes this was a matter of active rejection by the female,* but other times it was achieved through physical barriers (as in "look, but don't touch"), or by giving the males access only to, er . . . *decapitated* females (who obviously could not give active consent).† Afterward, each group of males, "mated" and "unmated," was given a choice between their usual mashed fruit and the same mash laced with 15 percent ethanol—an alcohol concentration comparable to that in a nice Chianti.[27]

Well, as it turns out, when male *Drosophila* fail to obtain "female companionship," they try to drown their sorrows. Yes, "unmated" males preferred the alcohol-spiked food, while "mated" males showed no preference at all. The authors of this paper were very thorough, using other controls to make reasonably sure that they were observing a real effect. What's more, when the rejected males were given

---

* Male flies in this group were only given access to "mated" females, while the other group got "virgin" females. "Mated" female flies are more likely to rebuff a male's overtures (saying they have a headache/are busy washing their hair that night, etc.).

† Male flies *must* obtain active consent from the female in order to engage in the sexual act; otherwise, they will not do a thing. In my mind, this is a not-so-subtle lesson that fruit fly "culture" teaches us: even in (some) insects, *consent matters.*

another chance at love and succeeded the second time around, their preference for the spiked food disappeared.*

Ah, the humble fruit fly ... Not so different from us after all.

As curious and significant as these behavioral results were, the researchers went further, securing deeper insight into these behaviors by looking at a specific biochemical process in their subjects. In vertebrate animals (like us), alcohol-related behaviors are closely related to a compound nicknamed neuropeptide Y (NPY for short).[28] Fruit flies have an analogous peptide, called neuropeptide F (NPF). The scientists hypothesized that NPF might play a role in ethanol-influenced behaviors in *Drosophila*. It is important to point out that, up until this research was done, neither NPY nor NPF had been associated with any type of alcohol-related *sexual* behavior. Yet, sure enough, Dr. Shohat-Ophir and her collaborators found that when male *Drosophila* were rejected by females or otherwise sexually deprived, their levels of NPF dropped; this was followed by the preference for alcohol-laced food. Based on this, the researchers hypothesized that activation/deactivation of NPF signaling—independent of any mating failure—would influence alcohol preference. When they tested their idea (in the absolute absence of female flies, mind you), they indeed found that if they reduced the activity of NPF in any way, the affected male flies "turned to alcohol," just as they would have if rejected by, or denied access to, the fairer fly sex. Conversely, activating NPF receptors made male flies indifferent to ethanol.[29]

As expected with this kind of study, the general gist (though not the actual details) caught the attention of the media, and pretty soon the hyperbolic headlines were flying (pardon the pun).[30] After all, sex—especially when mixed with alcohol—sells. Even though Dr.

---

* It is important to point out that alcohol affects female and male *Drosophila* differently. Males are more likely to get hyperactive, while in turn females tend to be more sensitive to alcohol's sedating effects. This difference seems to be sex-specific, since male and female fruit flies do not differ much in their build or weight (factors we might usually associate with alcohol resistance).

Shohat-Ophir and her collaborators were very careful to present and discuss their findings strictly in terms of *Drosophila*'s biology, some mainstream articles reporting on these findings made the direct leap from flies to people, in effect speculating that alcohol has the same effects in these two very different types of organisms. To be fair, this is not a completely untenable assumption, but if not backed by evidence, it is nothing more. Again, understandably, it is very hard to avoid anthropomorphizing when interpreting results observed in nonhuman organisms. For example, the Shohat-Ophir paper contained phrases like "ethanol intoxication" and "social experience of rejection," which were easily taken out of context for the sake of "exciting" headlines.[31] The flurry of press coverage rapidly encouraged the publication of scientific articles examining alternative explanations for the results shown in the Shohat-Ophir paper, including interpretations based on evolutionary and ecological insights. For instance, the attraction of "unmated" males to ethanol might have nothing to do with seeking solace, but rather could be related to the increased caloric content of alcohol improving fitness (and perhaps mating success in the future), or the fact that fermented fruits may be likely "gathering places" for other flies, including potential mates.[32]

The curious scientists kept digging, and in 2018, Dr. Shohat-Ophir (now the head of her own research group at Bar-Ilan University in Israel) and her collaborators published a paper reporting the discovery of a direct link in *Drosophila* between alcohol consumption and sexually derived physical pleasure.[33] In this study (led by Dr. Shir Zer-Krispil), the scientists examined specific molecular aspects of the reward system[34] in fruit flies, namely the role of a compound called *corazonin* (CRZ). Corazonin was named for the fact that it accelerates the heart rate in cockroaches,* but corazonin and corazonin-like compounds are

---

* More specifically, corazonin's name comes from *corazón*, the Spanish word for "heart." It is impossible for me not to note the coincidence of a chemical named after the heart that also affects an aspect of an organism's sexual function.

found in a wide (and I do mean *wide*) variety of invertebrate organisms, controlling many different functions. As we've said before, if something is found across many species in this way, it is bound to be important; otherwise, evolution would likely have weeded it out. And indeed, corazonin belongs to the larger family of gonadotropin-releasing hormones (GnRH), representatives of which are found in virtually every type of animal, invertebrate and vertebrate.[35]

The new research by Drs. Shohat-Ophir and Zer-Krispil showed, among other things, that male *Drosophila* ejaculated shortly after certain corazonin-producing neurons in their brains became active. Curiously, using behavioral evaluation methods similar to those from the 2012 experiment, the scientists found that, besides inducing ejaculation, the activation of these neurons eliminated the ethanol preference in "unmated" males. One straightforward interpretation of this observation is that the lack of physical pleasure induced by ejaculation is at least a factor triggering ethanol preference in *Drosophila* males.* You will recall that the team's earlier research showed that the levels of NPF in male flies influenced whether they preferred ethanol or not. To date, a link between corazonin and NPF has not yet been established, but I, for one, will be watching the relevant scientific databases very closely, as I fully expect that the tale of the lovelorn drunken flies will reach a happy conclusion sometime soon.

### Blurred Lines

Another alcohol-induced sexual behavior displayed by male fruit flies was described in a paper with the somewhat uninspiring title "Recurring Ethanol Exposure Induces Disinhibited Courtship in *Drosophila*."[36] This study, from the group of Dr. Kyung-An Han, then at Pennsylvania State University, essentially shows that male fruit flies become sex-crazed with repeated exposure to alcohol—but there's a

---

* Of course, it is next to impossible to know whether flies (or any other organism for that matter) experience pleasure as humans do; please see chapter three.

twist. The male flies became hypersexual, but not only toward female flies; many actively courted males with at least the same frequency. This particular behavior is called *intermale courtship*, and it is not unique to this experiment; when females are conspicuously absent from their environment, male fruit flies will actively and aggressively pursue other males in such a way that a group of these "hyperexcited" males forms a "conga line" a few flies long, mouth to . . . backside. An increase in this hypersexual behavior (conga line formation) was confirmed to be directly related to ethanol consumption because, simply stated, the greater the frequency of exposure to ethanol, the greater the incidence of the behavior. Repeating rounds of exposure increased the likelihood of intermale courtship such that, by the sixth round of ethanol exposure, roughly 40 percent of males engaged in the practice. The researchers traced this particular behavior to three molecular entities: a transcription factor, a transporter, and notably, the neurotransmitter dopamine, which as we saw earlier, plays a prominent role in addictive behaviors. In fact, dopamine appears to play a central role in this behavior as well, with mutations in the genes for the other two players (the transcription factor and the transporter) modulating the dopamine-dependent, ethanol-induced practice of intermale courtship.[37] Oddly, the age of the fly seems to have some influence as well—the older the male, the more "willing" he was to engage in intermale courtship.*

Another interesting observation reported by Dr. Han's group is that even though alcohol exposure induces enthusiastic courting behavior toward both males and females, when a female fruit fly

---

* In their paper, Dr. Han and her collaborators described yet another ingenious apparatus, like the aforementioned inebriometer, designed to measure ethanol-induced behavioral effects in *Drosophila*. Introducing *Flypub*: essentially a box with a transparent cover linked to a videotaping device and an open lower half through which to administer ethanol fumes. There is at least one other similar contraption in use as well—namely *FlyBar*, invented by Dr. Lisa C. Lyons of Florida State University and her collaborators to study how circadian rhythms are influenced by alcohol (more on circadian rhythms and flies soon).

signaled the equivalent of "Why not? Go ahead!" to a buzzed male, the eager suitor nevertheless frequently failed to perform. An article in the journal *Nature*[38] that reported on this research included an apropos quote from Shakespeare's *Macbeth*, in which a character remarks of alcoholic drink: "[It] provokes the desire, but it takes away the performance." Again, flies and people, not so different after all.

## The Unlikeliest Nanny

The main (practical) idea behind the phenomenon of sexual behavior is to beget offspring, and perhaps unsurprisingly, alcohol plays a role in this chapter of *Drosophila*'s biological life as well.

That *Drosophila*'s common name is *fruit fly* may have clued you in to the fact that fruit is the main source of calories for these organisms. We already know that ethanol can be toxic, and that because of this, many types of organisms that feed on fruit display some level of resistance to alcohol, enabling them to bypass its toxicity and use it instead as an energy source. Fruit flies are experts at this.[39] Researchers have even discovered interesting variations of ethanol metabolism and tolerance in flies of the same species coming from different geographical regions (where flies might be expected to encounter different fruits).[40] In other words, while alcohol resistance in *Drosophila* is largely a matter of genetics, there are also environmental factors at play.[41] At any rate, alcohol resistance in fruit flies is not limited to adults. Like many fly species, *Drosophila* females deposit their eggs on food—in *Drosophila*'s case, this means (sometimes fermented) fruit. And so, evolution has endowed *Drosophila* larvae and their eggs with resistance to alcohol concentrations that would be harmful to other invertebrate species. This is advantageous to the hatchlings, since, after all, fermented fruit may well be the first food source they consume. And Mama Fly takes full advantage of her offspring's resistance to alcohol, using it as a defensive weapon to protect her babies.

Just as we have "legitimate" (i.e., non-recreational) uses for ethanol, namely as a disinfectant and component of various medications, fruit flies display remarkably similar behaviors that take advantage of ethanol's chemical properties. One of these is the use of ethanol in antiparasitic defenses. As we saw earlier, the continuous evolutionary war between parasites and hosts is a powerful source of selective pressure in many species, which may evolve chemical defensive strategies in response. In order to defend their eggs and larvae against parasites and parasitoids*—including certain species of wasps that deposit their eggs in *Drosophila* larvae—several *Drosophila* species have evolved to use alcohol as a defense. Fruit flies generally display higher resistance to ethanol toxicity than their parasites (more specifically, in the case of the parasitic wasps, the larval flies display a higher tolerance to ethanol than the wasp larvae). Female fruit flies can take advantage of this by choosing to deposit their eggs (which are also ethanol-resistant) in fermenting fruit, which will naturally contain significant concentrations of ethanol. In general, the higher the ethanol concentration of the fruit, the more attractive it is to female *Drosophila* for egg-laying purposes. Remarkably, while female fruit flies do not always lay eggs in alcohol-laden fruit, they almost invariably do so *when they detect the presence of a parasitic wasp*.[42] Scientists have already identified the genes responsible for this cunning and sensible behavior in *Drosophila*; moreover, they discovered that NPF (which you will recall from above) plays a role in it. For example, if NPF levels are reduced, this egg-laying behavior is increased, and vice versa.

If you think about it, this fruit fly behavior is roughly akin to booby-trapping a nursery with chemical weapons (ones that won't harm baby, of course). As an admittedly nervous parent, I think this is a pretty cool idea. Alas, all my children are adults now and I must

---

* In brief, the main difference between the two is that a *parasite* "lives" with its host on a more or less permanent basis. On the other hand, a *parasitoid* has an adult stage that lives independently, needing the host only in its larval stage. This nuanced difference does not matter that much for our conversation, but I tend to be a stickler for details.

reluctantly admit that they probably do not need such an extreme protection strategy. Oh well—there's always my future grandchildren!

## TIME FLIES, OR IT'S FIVE O'CLOCK SOMEWHERE

Like virtually every part of the physical universe,* biological life is subject to the passage of time. Thus it is hardly surprising that evolution has endowed organisms with the capacity to measure and sense time and to use this physiological capacity in various ways to optimize survival. These time-related biological activities collectively fall under the umbrella of a scientific discipline called *chronobiology*, which studies the biological rhythms that control virtually every aspect of an organism's physiology in virtually every single class of organism.[43] In general, by virtue of our planet's twenty-four-hour solar day, these biological rhythms organize themselves in time periods roughly twenty-four hours long (the actual length ranges between twenty-two and twenty-five hours depending on the organism, though it can even vary between different types of cells in the same organism). Because these rhythms are not exactly twenty-four hours long (or twenty-three hours and fifty-six minutes, the actual—though still-approximate—length of a day on Earth), they are called *circadian*, meaning "close to a day" or, if you prefer, "a day-ish." The control of circadian rhythms is managed by biochemical systems fittingly called *biological clocks*, which are present in every single organism known so far, from single-celled species to whales. Biological clocks allow organisms to anticipate environmental changes independently from environmental signals. For example, evolution tends to favor the survival of an organism that wakes up a little before the sun actually comes out (the proverbial early bird and all that). Insights from

---

* With the possible exceptions of exotic places like the interior of black holes and places with a preponderance of dark matter and dark energy, of which we know very little. .

chronobiology are being applied to a variety of phenomena relevant to our daily lives, from reproductive cycles to the diagnosis and treatment of medical conditions that result in sleep disturbances. Several types of invertebrates and *Drosophila* in particular are considered useful models to study the physiology of sleep and memory as a function of circadian rhythms, with the objective of better understanding problems related to unusual work schedules, jet lag, and so on.[44] This research can lead to the development of novel drug therapies, as many chemicals are involved in (and can affect) an organism's chronobiological processes.[45] What's more, we now know that the effectiveness of drugs can vary according to the time of day, giving rise to the concepts of *chronotherapy* and *chronopharmacology*, which aim to take these factors into account when administering and studying medications and other substances, even nutrients.[46]

An interesting aspect of the chronobiology of the fly involves the interplay between time of day and the effects of alcohol and other abused drugs. In *Drosophila*, the effects of alcohol display sensitivity to circadian rhythms: for example, flies show increased tolerance to ethanol at the end of the day (meaning they need more of it to experience psychoactive effects). Conversely, lesser amounts of ethanol are needed to get a fly "buzzed" in the morning.* One related behavioral measure that is influenced by circadian rhythms is the loss-of-righting reflex. Briefly, this measures the amount of time a sedated fly needs to get "back on its feet" after being gently turned upside down (essentially a measurement of the recovery of motor function).[47] *Drosophila* is being developed as a model to study the effects of biological clocks in the context of other abused drugs in a variety of organisms, including humans.[48] Just like aging humans, for example, aging flies react differently to drugs. They simultaneously experience increased sensitivity to alcohol and extended recovery of function times, both of which are modulated by biological clocks.[49] I

---

* The phrase "it is too early to start drinking" comes to mind.

am sure that chronobiology, and our good friend *Drosophila*, still have many surprises in store for us . . . only time will tell.[50]

## SPIDERS ON DRUGS

One of the most compelling characteristics of science is that nobody has the foggiest idea of when some "routine" observation will bring an attention-grabbing, even significant, discovery. These are quite unpredictable events, and this is precisely the reason that fundamental research—that is, research with no apparent immediate application—is so important. And when scientists with different interests and expertise interact, oftentimes their combined knowledge coalesces in unexpected ways, generating insights that otherwise would most likely never have seen the light of day. The story of the discovery that specific drugs induce specific effects on the web-building behavior of spiders is a perfect example of the importance of both serendipity and collaboration in science. This is a tale of two scientists who found themselves frustrated by disappointing results in their respective lines of research. Driven by their curiosity about nature, they decided to combine forces, and nature rewarded their determination. But I am getting a little ahead of myself. Let's begin this story with the question I am sure you are already asking: Who in their right mind would think of giving drugs to spiders in the first place? The short answer, of course, is, "Scientists; that's who."

It all started in 1948 at the University of Tübingen, Germany, with a group of zoologists led by Dr. Hans Peters. The Peters group wanted to film spiders building webs as part of a nature documentary. Alas, as frequently happens with experimental animals, the spiders had minds of their own and cared not at all about what the scientists were trying to accomplish. This particular spider species (*Zygiella x-notata*[51]) is an orb-weaver spider, which—like all spiders—has the ability to produce silk[52] in order to weave exquisite, delicate, and yet sturdy and fully functional webs with the main purpose of capturing prey.

The fact that *Zygiella* was a nocturnal spider presented a big problem for Dr. Peters's group, who collectively grew increasingly frustrated by their inability to catch the spiders in the act of building their intricate webs. The tired scientists watched the spiders for as long as they could, staying up later and later, but eventually they would turn in for the night, only to find beautiful, fully constructed webs the following morning. In desperation, they turned to their friendly neighborhood pharmacologist,* Dr. Peter Witt, who would later write in a popular article about the affair: "The members of the zoology department were dead tired . . ."[53]

What the zoologists did not know was that Dr. Witt was also frustrated with his own research. A physician before he was a pharmacologist, he had a keen interest in the effects of psychoactive drugs on humans, but while his human subjects were more cooperative than Dr. Peters's spiders, they presented problems nonetheless. Most indeed reported psychoactive effects in response to the drugs Dr. Witt administered, but it was often difficult to differentiate these effects between drugs—for instance, cannabis versus morphine. Drowsiness looks like drowsiness regardless of the drug that causes it, meaning that drowsiness was likely a "generic" effect of many narcotics and other similar drugs, and told Dr. Witt relatively little. As a scientist, I completely understand his frustration. After all, you cannot write (or at least, you are unlikely to publish) a scientific paper that says merely: "We observed that all the tested drugs induced drowsiness in humans. The end."

Anyhow, to continue our story, the zoologist (Peters) asked the pharmacologist (Witt) whether he knew of any drug that could "confuse" the spiders, with the purpose of "tricking" them into weaving their webs in broad daylight so that the research team could more

---

* You can never go wrong with a pharmacologist.

easily capture their activities on film.* Dr. Witt was more than willing to help, but told them honestly that he knew very little about spiders and their web-building behaviors and even less about the effects of drugs on these animals—he did not even know whether the nervous system of a spider would display *any* reaction upon exposure to psychoactive substances. This caveat in place, the pharmacologist told the zoologists that they were welcome to drug samples for testing, which they happily took (not themselves—they gave the drugs to the spiders; we'll go over how they did so in a bit).

According to Dr. Witt, the zoologists returned the next morning with bad news: the spiders were still only weaving their webs at night. And there was worse news still: the webs of the drug-dosed spiders did not "look right." Depending on the drug given, some spiders made webs with weird or incomplete shapes. Other spiders seemed "uninterested" in building webs altogether. But as is often the case in life, one person's trash is another's treasure, and what annoyed the zoologist excited the pharmacologist greatly. You see, what Dr. Witt realized—and eagerly shared with Dr. Peters—was that by giving drugs to spiders and observing how they subsequently built their webs, one could make inferences about how drugs affected their behavior more generally. Dr. Peters quickly grasped the potential and became excited himself.

The spiderweb-building behavioral model was especially attractive as many aspects of this behavior were already understood by Dr. Peters's group (and by spider zoologists in general).[54] For example, it was known that web building is an innate behavior; a mother spider does not teach her offspring how to build a web, meaning that this ability is instinctual (in other words, genetically encoded). They also knew that the hungrier a spider is, the harder and faster she works on her web,† and that if a spider's web is destroyed, it invariably builds

---

* They did not realize this at the time, but essentially what they were hoping to do was pharmacologically modify the spiders' circadian clocks.
† In the overwhelming majority of cases, although male spiders can produce silk, it is the female spiders that actually build webs.

another one by the next day. And by virtue of their work, they had become very familiar with the general patterns and techniques that various spider species used in web construction. So, armed with this information and Dr. Witt's pharmacological expertise, the collaborators set out to study the effects of drugs on spiders. This fruitful collaboration would last for several years, producing many insights and quite a few scientific papers.

A critical factor on which their studies depended, of course, was the ability to give various drugs to spiders in the first place, which is not as easy as it might sound. I confess that at first I assumed they injected the spiders with a solution, and I marveled at the impressive manual dexterity of our protagonists, given that *Zygiella x-notata* is not a big spider (females average about 11 mm long)—what fine needles they must have needed, and how delicate and precise their technique to avoid hurting the spiders in the process!*

As it turns out, the scientists did *not* actually inject the spiders with drugs. Their initial method was elegantly simpler: they spiked the spiders' water. Specifically, they dissolved the drugs in sugar water (to disguise the bitter flavor common to many alkaloid drugs) and allowed the spiders to drink the solution at will. Other times, they used a syringe to "feed" the liquid solution to the spiders, which said spiders apparently imbibed with gusto. I must admit that I am somewhat surprised this strategy worked at all. For one thing, spiders are carnivorous, with only one known exception so far among about forty thousand known spider species.[55] Thus, one would not expect them to be attracted to sweet foods (and I am not even sure they would perceive sweetness as mammals do). Secondly, if I were to see a giant coming toward me with what, from my perspective, is for all intents and purposes a deadly spear, I would not be in the mood for a drink, to say the least. Perhaps this is why they did not have as much success as they were hoping with this strategy—but they refined their methods,

---

* Ah, that blissful state of being unburdened by knowledge!

coming up with various ingenious and more "natural" ways of dosing their subjects. One clever iteration of their experimental design involved injecting the decapitated body of a fly with the drug-laced sugar solution and offering that to the spider. But even then, they ran into some procedural difficulties. Namely, when the researchers put this enticing lunch in front of a spider, it was flatly ignored. After pondering possible reasons a spider might reject a free meal, they realized that they'd overlooked a crucial detail: the type of spiders they were studying are not particularly visual creatures. Rather, they react to the vibrations in their webs caused by trapped insects; these vibrations trigger the spiders' "attack response" and guide them to their meal. The scientists' solution was ingenious: they put the spiked fly body on the spider's web and made the web vibrate using a tuning fork with a vibrating frequency similar to the one induced by a live fly.* Upon sensing the vibration, the spider pounced on the headless body and happily fed on it. This strategy paid off handsomely, allowing the scientists to gather enough evidence to identify patterns in the kinds of webs built by drugged spiders. Ironically, they were never able to film the spiders "in action," as it were. The zoologists would be ready to film, every piece of equipment set up, and as long as they were in the room, the spiders would not build anything. They began to build their webs only when people left, and to this day, nobody knows why. It is unclear whether the scientists conducted their experiments in such a way as to minimize "distractions" for their subjects. These animals display exquisite sensitivity to vibrations, and perhaps

---

* Incidentally, this is a perfect illustration of the importance of looking as far back as you can when researching the published literature on your subject of interest. The spider response to the vibrations of a tuning fork was first described and published in the 1880s. Please see Boys (1880) and Barrows (1915). As the biologist Holger Valdemar Brøndsted said in his book *Planarian Regeneration* (published in 1969): "When you have done an experiment, then go through the [scientific] literature until you have found that your experiment has already been performed by others. Only then can you be sure that you have been through the entire literature."

these were enough to inhibit the spiders, who apparently valued their privacy.

Anyway, once the methodological wrinkles were ironed out, the researchers were indeed able to observe and infer specific behavioral effects in the spiders upon drug consumption as evidenced by their misshapen webs. The types of "mistakes" that the spiders made during the web-building process seemed to consistently correlate with the type of drug given. The scientists also observed that if they gave too much of any drug to a spider, the animal exhibited very limited movement (drowsiness?), which in retrospect is not surprising, as practically any substance given in very high amounts (especially a drug) is likely to induce undesirable effects such as toxicity, as we saw in chapter two (remember, the dose makes the poison). In time, the researchers figured out the optimal doses needed to observe "non-drowsiness-related" behaviors. Depending on the drug tested, they observed differences in the size of the webs, their specific design and organization, and whether the spider completed the web at all.[56]

Dr. Witt and his collaborators also painstakingly measured many aspects of the architecture of the webs, such as the number and thickness of the threads and the angles between them as well as their general shapes under normal conditions and "under the influence," hoping to quantify these properties and use them for diagnostic purposes. More precisely, Dr. Witt's hope was to develop a bioassay to test for the presence of abused drugs in human fluids (plasma, urine, etc.). Imagine you are taking a standard preemployment drug test—except that the sample you provide is given to a spider, and whether and what drugs are present is determined based on the webs the spider builds upon exposure. This is not as odd as you may think, as at the time many tests for the presence of chemicals involved exposing an animal to a sample (probably the most famous example of this is the rabbit-based pregnancy tests then in use). But while Dr. Witt's idea was a sound one, it never panned out. For one thing, as interesting as it was to observe what drugs did to a spider's behavior, the information that could be obtained from such experiments was limited; at the time, very little was known about the neurochemistry of spiders and how it influenced their behavior. Nonetheless, Dr. Witt's research inspired other scientists to pursue similar endeavors. In the 1990s, scientists from NASA replicated several of these kinds of experiments using spiders, with the objective of assessing the toxicity of various chemicals. They did this by analyzing the shapes of webs built after spiders were exposed to compounds like caffeine, cannabinoids, chloral hydrate, and other substances (Figure 5.2).[57]

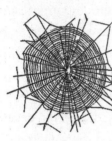  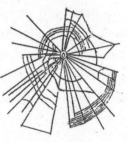

No drug          Tetrahydrocannabinol          Amphetamine
                        (THC)

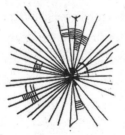 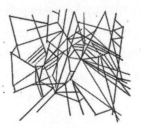

Lysergic acid                Caffeine                Chloral hydrate
diethylamide (LSD)

Figure 5.2. Representation of spiderwebs made upon exposure
to selected drugs. Drawn by my way better and prettier half, Mrs.
Elizabeth Rivera. Based on NASA Tech Briefs, April, 1995.

The serendipitous collaboration between Drs. Peters and Witt was one of the earliest instances of "lower animals" being given psychoactive drugs on purpose in order to see what happened (sci-entifically, at least). Their work is important because it paved the way for the systematic exploration of the pharmacology of abused drugs in nonmammalian animals, which is why I chose to tell the story in some detail here. Studying the antics of spiders on drugs beautifully illustrates the relationship between drug exposure and behavior in general. Despite their dubious reputation as "heebie-jeebie genera-tors," arachnids have a lot to teach us, and it would not be a stretch to say that we can see ourselves in these invertebrate examples of apparent "intoxication."

## HALLUCINATING SEA SLUGS

Along with the lack of knowledge at the time regarding spider neuro-chemistry, another factor that prevented the scientists we met above from making more concrete interpretations of their observations was that they had no way to know how the spiders experienced the effects of the drugs they were given; they only had the shapes of their webs to work with. This is a problem for all scientists working in animal behavior: we cannot ask our subjects how they "feel." And, of course, this is especially a problem when it comes to evaluating more "subjec-tive" drug effects, like hallucinations.

We broadly define hallucinations as subjective perceptions (visual, auditory, olfactory, tactile, gustatory, and so on) with no real cause. This is an admittedly vertebrate-biased (and mainly human-biased) definition; as we discussed in an earlier chapter, there are "senses" present in other organisms for which we humans have no frame of reference. How would we understand what a bat feels when halluci-nating an echolocation, for example? The reality is that we can tell with relative certainty whether an organism is hallucinating only in the case of humans, with whom we can generally communicate about experiences. Sometimes, given what we know about our own neurochemistry and the properties of various substances, we can get a pretty good idea when it comes to other vertebrates: if, after being exposed to a medication or a toxin, your dog is growling at what appears to be nothing, it is a safe bet that your animal companion is experiencing a hallucination. Then again, we cannot be *sure* of exactly why your dog is growling. There is even greater difficulty, as you might imagine, in interpreting certain behaviors as hallucinations in the case of "lower organisms" like invertebrates. How can we tell whether a cockroach is hallucinating? Difficulty notwithstanding, some quite intriguing research nonetheless claims to demonstrate hallucinations in a certain invertebrate—namely, a sea slug.

*Tritonia diomedea* is a species of marine slug (specifically a *nudi-branch*, most species of which are quite beautiful; we met some especially clever ones in chapter three) widely distributed throughout the proverbial seven seas. Sea slugs have played a significant role in the historical development of several scientific disciplines, the neurosciences in particular. In fact, they've garnered at least one Nobel Prize—awarded not to a sea slug but to Dr. Eric Kandel, for his research on neurobiological mechanisms of memory in sea slugs.[58] As a part of this distinguished tradition, a recent paper on *T. diomedea* reported what is possibly the first unambiguous evidence of hallucinatory behavior in an invertebrate, seemingly induced by exposure to amphetamines, a well-known category of abused drugs.[59] Obviously, nobody asked these slugs: "What do you see?" So what made the scientists involved in this research think that these animals were hallucinating? And what possessed them to consider sea slugs as candidates for research into an experience as subjective as hallucinations?

Amusingly, Dr. Anne H. Lee and her collaborators at the Rosalind Franklin University of Medicine and Science in Chicago did not set out to study hallucinations in *T. diomedea* (again, while sea slugs may look like something you'd *see* in a hallucination, who thinks of sea slugs hallucinating?). These scientists simply wanted to study the effects of amphetamines in these organisms: pure, curiosity-driven research. A common behavior in many sea slugs, *T. diomedea* among them, is the "escape swim response," which is fairly self-explanatory. You can elicit this response with even the gentlest touch; basically, the slightest tactile stimulation triggers the marine version of "run for your life!" behavior. In general, because this behavior results from touch stimuli, if you leave the slug alone, it will, well, not do this. But when these intrepid researchers exposed the slugs to amphetamines, they observed something quite unexpected. After administering the amphetamine injection, they left their experimental subjects well alone for a period of time, carefully observing and monitoring their behavior—and saw many of the slugs "fleeing" without being touched. The researchers knew that amphetamines can cause hallucinations in humans, and therefore hypothesized that the slugs were experiencing drug-induced hallucinations as well. Naturally, they did not stop there. They did further experiments, showing that when dissected *T. diomedea* brains (just the brains, mind you, not attached to a body) were exposed to amphetamines,* a neuronal pathway that controls the activation of movement in these slugs became active. Better still, they were able to trace this activation to a particular type of cell that detects tactile stimuli. Taken together, their results seem to indicate that the nervous system of a humble sea slug is indeed capable of at least hallucinatory-like behavior. I am quite confident that these slugs will not be the only invertebrates found to experience hallucinations, once we figure out how to ask the right questions (and interpret the answers). But for the time being, in light of the lack of reliable animal models to study hallucinations, these results in *T. diomedea* are of particular importance, and I predict an upswell of interest in our sluggish friends.

---

* Brain in a vat, anyone?

## DRUGGED FLATWORMS

This probably does not come as a surprise by this point, but I find flatworms utterly bewitching, particularly the type of flatworms known as *planarians* (Figure 5.3).

Figure 5.3. Artistic representation of a lovable
planarian. Drawn by Ms. Chelsea Linaeve.

I've already mentioned their historical importance to science, and though their popularity has been overshadowed by that of the fruit fly as a model for genetic study, I nonetheless have many sound, scientific reasons to find them interesting, particularly from an evolutionary perspective. Planarians are distributed widely throughout the world in virtually every possible habitat; there are terrestrial, marine, and freshwater species—all of them predators. These organisms represent some of the earliest beings to evolve a head with sensory organs like eyes, and more importantly, despite their simplicity, the planarian head houses a relatively sophisticated brain.[60] Another astonishing ability that many planarian species have mastered is the regeneration of body parts and organs—that relatively sophisticated brain included.[61] I also happen to think these worms are rather cute, but recognize that this is not a widely held opinion, and anyway not relevant to the subject I want to discuss here, which is the relationship between freshwater planarians and abused drugs.

Though there are reports dating back to the 1800s of naturalists giving cocaine or nicotine to a variety of flatworm species in order to slow them down and make them easier to study, systematic research into abused drugs using planarians as subjects only began about thirty years ago.[62] Since then, however, the popularity of the planarian model (in particular freshwater planarians, which are the best known)[*] as applied to the pharmacological sciences has grown steadily. Over the last thirty years or so, scientists have given virtually every known type of abused drug to planarians—including ethanol, nicotine, cocaine, amphetamines, opiates, cannabinoids, and so on—and the effects of those drugs on this organism have proven to be remarkably similar to those caused in vertebrate organisms, including humans.[63] One example of this is withdrawal-like (sometimes termed "drug-dependent") behaviors. If you put some cocaine in the water planarians live in, leave the compound there for a while, and then transfer the worms to cocaine-free water, these worms will express behaviors that closely resemble withdrawal behaviors in mammals and other "higher" organisms, including humans. Some worms will get "the shakes," some will swim while performing "corkscrew" movements, and yet others will swing and twist their heads, among other behaviors.

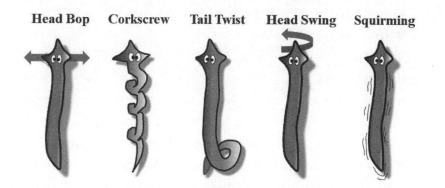

*Figure 5.4. Planarian withdrawal-like behaviors in response to cocaine exposure. Drawn by Dr. Robert B. Raffa.*

---

[*] From this point on, when I say "planarians," you may assume I mean "freshwater planarians."

Significantly, the frequency of these behaviors is directly proportional to the amount of drug administered, and after a while, the worms gradually return to their placid lives.[64] These two latter aspects strongly argue for a pharmacological cause.*

Please note that I am not simply reporting on these behaviors; I have seen them firsthand—as mentioned before, I am one of those scientists who uses planarians in their research. One of the research objectives of my laboratory is to identify compounds that might be useful in treating drug-induced toxicity (we have actually found a few compounds of interest, which we continue to study),[65] and I've recently become interested in the pharmacology of regeneration—which includes the search for compounds that might enhance the regeneration process in "higher" organisms, including us.[66] My laboratory is only one of a growing number of groups worldwide taking advantage of the usefulness and versatility of planarians, and it seems the planarian model is here to stay. Despite their status as "lowly" invertebrates, the fact is that when exposed to abused drugs, planarians display many common and well-understood drug-related effects, including tolerance, dependence, and "anxiety-like" behaviors, among others.[67]

Wait—*anxiety-like behaviors?*

## Anxious Flatworms

There is little dispute that we live in an age of anxiety. The reasons for this are complex, to say the least, but at least one factor is that most of us live in an environment that exposes us to almost constant stress, a situation that was never biologically meant to be.[68] As a direct consequence, anxiety-related disorders (including generalized anxiety disorder, post-traumatic stress disorder (PTSD), and others) are an actual epidemic in our society. The current animal models used to study anxiety disorders are almost invariably

---

* Please refer to "The Philosophy of Pharmacology" in chapter two.

mammals—the usual suspects, namely rats and mice[69]—and there is a clear need for alternate models, since large-scale experiments with rats and mice are costly and sometimes impractical. Invertebrates offer an attractive alternative. And one particular characteristic that planarians share with rats and mice has made them a front-runner in this area.

Evaluating light/dark preference in rodents is one of the most widely used behavioral tests when studying anxiety-like disorders. Rats and mice have to be "shy" organisms—after all, they are not too big, they can't fly, they are not venomous, and they are preyed upon by virtually every small- to child-sized predator out there (domestic cats, for instance). Hiding, preferably in the dark, is one of their chief defensive strategies. Thus, an "anxious" rat will tend to spend more time in a dark place than in an illuminated place. Scientists have found that they can manipulate this behavior by giving a rat some of the very same antidepressants that work in humans.

If you were surprised to learn that we can treat "rat anxiety" with antidepressants, you will be amazed to learn that planarians react in a very similar way. Light/dark preference experiments and similar tests are now being conducted in planarians to evaluate whether these organisms might be useful models for the study of anxiety disorders.[70]

As a mental exercise, please put yourself into the metaphorical shoes of a planarian. You are small, you are not particularly fast, you are definitively not poisonous, and you are the proud owner of a fragile and squishy body. What do you do for protection? You hide. That's why planarians prefer dark places—generally beneath rocks and so on. The recognition of this fact allows scientists to design experiments using the planarians' tendency to hide as a surrogate for anxiety-like behavior in general. This strategy seems to be a fruitful one; recent work relying on this paradigm showed that planarians' anxiety-like behaviors are significantly reduced by an antidepressant (fluoxetine, often branded as Prozac) widely used to treat anxiety.[71] This latest demonstration of the usefulness of planarians in the study of phar-macological phenomena adds to a robust body of research built by scientists over the past few decades, showing the outsize significance of these tiny worms in terms of what they can help us learn about drugs and their effects.

## GRUMPY OCTOPI ON ECSTASY

Some of the most interesting and unusual organisms on this planet, octopi (Figure 5.5)* are simply full of surprises. Despite possessing a nervous system architecture *markedly* different from that possessed by "brainy" animals like vertebrates, they display undeniable cogni-tive capabilities. These include but are not limited to curiosity and intelligence, the ability to solve mazes, the ability to solve complex problems, to learn by observing, and even to use tools. They have been said to express individual "personalities," and many species have mas-tered an uncanny strategy of hiding by blending into their environ-ment in a way that goes far beyond something as simple as changing color—this sophisticated camouflage behavior includes the ability to change texture, as well as to imitate dangerous animals to intimidate potential predators.[72]

---

* I love learning new things. I'd long thought that the correct plural form of *octopus* was *octopuses*, not *octopi*: the "common wisdom" was that the etymological root of the word "octopus" came from the Greek language; therefore, the true plural would be "octopode" (with "octopuses" an acceptable variant), but because the language of the natural sciences at the time the organisms were first described was Latin, people understandably assumed that the plural form was "octopi." As it turns out, however, this narrative is incorrect! I am reading a delightful book, *Monarchs of the Sea*, by Dr. Danna Staaf, in which she untangles this issue right in the first chapter. Bottom line: feel free to use octopi, octopuses, or octopodes, according to your preference. I am partial to octopi. It sounds cooler.

*Figure 5.5. Representation of an atypically happy octopus. In this case, the octopus is happy because it's his personality, not because of drugs. This illustration is the kind gift of Dr. Danna Staaf.*

Cephalopods in general (a group that also includes squid and cuttlefish), and octopi in particular, are widely considered the smartest invertebrates,[73] and their evident intelligence rightly fascinates us for a variety of reasons. Most basically (though erroneously), we do not tend to associate intelligence with invertebrates, and encountering a "spineless" (a word oftentimes used as a pejorative) organism clever enough to solve puzzles, well, puzzles us, to put it mildly. In fact, octopi seem so incongruously far from us physiologically while being so similar cognitively that Dr. Peter Godfrey-Smith, a professor in the School of History and Philosophy of Science at the University of Sydney, opines that octopi are "probably the closest we will come to meeting an intelligent alien."[74] In 2018, some scientists took this idea a little too far, speculating—among other unrelated notions—that octopi are not simply alien to us, as in weird, but actually, literally, *aliens.*[75] As you can imagine, this extraterrestrial interpretation was (and still is) soundly rejected by the vast majority of scientists (myself included), in no small part because of the ample available evidence of cephalopod evolution on good old planet Earth. Some of this evidence includes the sequencing of the genome of an octopus species that unambiguously demonstrates that octopi and their kind, weird as they are, evolved right here.[76]

One result of advances in the study of genomes is that scientists can use this data to explore the similarities and differences between groups of organisms. When the genome of *Octopus bimaculoides* was sequenced and compared to the human genome, it became evident that these invertebrates had some striking neurochemical similarities to humans. One such similarity that caught the attention of Dr. Gül Dölen at Johns Hopkins University was that the serotonin system in this species of octopus seems to closely parallel the serotonin system in humans.[77] Specifically, certain gene sequences encoding the proteins that interact with serotonin were very similar in *O. bimaculoides* and in humans; in fact, some of these gene segments were identical in our two species. This was intriguing, in part, because in mammals (including humans), there seems to be a relationship between the serotonin system and social tendencies.

Before we go any further, I must tell you that despite their intelligence (or perhaps because of it), octopi are notorious curmudgeons. With a few exceptions—that are only now beginning to be studied properly[78]—they do not tend to be social animals. This is something of an understatement, actually: octopi in captivity are housed in separate enclosures because of the high likelihood that they would otherwise kill or eat one another. When the, er, proximity necessary to the process of sexual reproduction requires them to engage in some degree of social interaction, they do not seem to find it enjoyable (or like it at all, for that matter); in fact, mating is oftentimes a risky, even lethal, proposition for both sexes. Humans, on the other hand, are known to be very social indeed, making this particular genomic similarity both perplexing and exciting.

Dr. Dölen's group was not primarily concerned with octopi but with social behavior more generally, and they had done a great deal of research using the drug MDMA

(3,4-Methylenedioxymethamphetamine, in case you are curious—Figure 5.6), better known as *ecstasy*. Recently, this drug has been considered as a possible treatment for certain neuropsychiatric conditions such as anxiety, depression, and post-traumatic stress disorder,[79] but it is more familiar in its use as a recreational drug. It is a well-understood substance with effects on (among other things) perception, mood, and affective behaviors. More to the point, it seems to have "prosocial" effects—in humans it increases feelings of "closeness to others," and has been shown to increase socialization in mammals more generally.[80] MDMA works by binding to serotonin transporter sites; the protein for this serotonin transporter was among those that genomic analysis revealed to be very similar in both *Octopus bimaculoides* and in humans.

I think you can see where this is going.

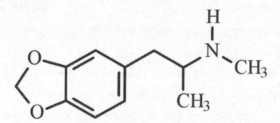

Figure 5.6. MDMA (pretty little thing; it reminds me of a tadpole). Drawn by the author.

Dr. Dölen teamed up with Dr. Eric Edsinger, a marine biologist then at the Woods Hole Oceanographic Institution. Upon running preliminary studies to assess the baseline behavior of their animal subjects, the scientists were surprised to discover hints of social interest—at least under specific circumstances. Basically, the test works by putting a subject octopus in the middle chamber of a three-chambered tank. On one side is a chamber containing another octopus (who is enclosed in a sort of basket); on the other is a chamber containing an interesting object, such as (and I am not kidding you here) a Star Wars figurine. When the other octopus was a male, the subject acted as expected, spending most of its time in the toy chamber and generally staying well away from the chamber with the other octopus. When that other octopus was a female, however, the scientists were surprised to find that their subjects spent most of their time in the chamber with her—this was true whether the subject octopus was male or female, and didn't seem to be related to size or any other obvious characteristic.

The scientists reasoned that to test the effects of MDMA on social behaviors, they would do best to use only males in the "social option" chamber—their evidence would be strongest if they pitted the drug against the variety of socialization their subjects had shown a distinct aversion to. And to minimize discomfort, they did not inject the MDMA directly into the octopi, but instead dissolved it in their water. This was probably for the best—I do not feel very social when poked with needles, do you? Plus, have you ever tried to stick a needle into a slippery, strong, and quite intelligent animal who just so happens to have *eight arms*?

Thrillingly, the effect was undeniable. Even when the octopi had shown interest in their fellow creature previously, it had been tentative, with any exploratory touch usually involving only a single extended tentacle. Dosed with ecstasy, however, the octopi not only spent far more time in the "social" chamber than they ever had before (even when the chamber had contained a female), but their behavior once in the chamber was transformed: they basically wrapped themselves around the other creature (the better to explore it). That octopi significantly increased their socialization when exposed to the drug at doses comparable to typical human consumption provided the first pharmacological evidence of parallels between behavioral circuits related to socialization in cephalopods and in humans. This opens the door to any number of exciting research possibilities, including into aspects of the evolution of behavior. In other words, these results could help us better understand how we humans (and others) evolved to be social creatures in the first place.

We have barely scratched the surface of psychoactive-induced behaviors in "little minds," but this book cannot go on forever. We've ended this chapter with a "brainy" invertebrate, one whose reactions might be taken to mimic more faithfully those in humans by virtue of its higher cognitive capacities, but I hope you were properly amazed at the markedly similar behaviors displayed even by organisms as different as planarians and people. Perhaps your curiosity has been piqued enough to inspire you to delve more deeply into the subject than we can go here—at the very least, I expect that the next time you see a bee, a spider, or any other of our spineless planetary cohabitants, you'll look at it a little differently.

# CHAPTER 6

# BIGGER MINDS ON DRUGS

*Folks ... please don't flush your drugs, m'kay?*
—POLICE CHIEF OF LORETTO, TENNESSEE, ON THE
LORETTO POLICE DEPARTMENT FACEBOOK PAGE

*As the LSD effect came on, forty minutes after the injection of
100 mcg, the dolphin came over to me. She had not approached
me before. She stayed still in the tank with one eye out of
water looking me in the eye for ten minutes without moving.*
—DR. JOHN C. LILLY, "DOLPHIN-
HUMAN RELATION AND LSD-25"

*Flapping their ears like a frustrated Dumbo ...*
—DR. RONALD K. SIEGEL, *INTOXICATION: THE
UNIVERSAL DRIVE FOR MIND-ALTERING SUBSTANCES*

In the previous chapter, we heard some stories from the world of
invertebrates, discovering further evidence of the biological and
chemical connections between us. This chapter moves on to the
vertebrate realm—from bees to birds, so to speak—and gives us a
chance to deepen our understanding of more "familiar" animals, even

some creatures we consider friends. You might expect to find even stronger similarities here among the bigger minds, and it is true that the animals in this chapter share aspects of physiology that make their behavioral responses to drugs easier to relate to our own. However, as we are about to be reminded once again, a spine isn't everything.

As we have seen, examples of species using substances for medicinal or recreational purposes can be found in virtually every type of organism, from invertebrates to our fellow primates. You will notice that I said "virtually." One of the oddities I discovered when researching this book is that there are almost no documented cases of reptiles,[1] amphibians, or fish* engaging in either zoopharmacognosy or purposeful intoxication with psychoactive substances. In fact, I was able to find only one: the Indian garden lizard (*Calotes versicolor*) apparently consuming the leaves of a certain plant, the holy basil (*Ocimum sanctum*), for medicinal purposes.[2] Of course, there is also the reference to opium-loving lizards in the Jean Cocteau quote at the beginning of chapter one, but there does not seem to be any scientific documentation of this affinity.

Fish and amphibians are the least understood in this regard mostly because of the inherent difficulties in studying their behavior in their natural underwater environments. However, the fact that we haven't been able to document these animals engaging in medicinal or recreational substance consumption doesn't mean they don't do so; in fact, I am confident we will find evidence that they do. My reason for this is the fact that there are a prodigious number of marine and presumably freshwater organisms that themselves produce a series of fascinating chemicals whose functions are, at the moment, grossly understudied. Many of these chemicals are previously unknown alkaloids, and others display relatively "unexplored" structural and

---

* If you are wondering about the dolphins mentioned in the title of this book, remember that, while fish-esque, they are in fact mammals. I promise to tell their story later in this chapter!

functional features.[3] Incidentally, some of my favorite products of marine pharmacology belong to a family of chemicals collectively called *cembranoids*, a family regarded as a promising possible source of novel therapeutic compounds.

## MYSTERIOUS CEMBRANOIDS

Cembranoids are compounds formed by a 14-carbon (or "cembrene") ring (Figure 6.1). They have been isolated from plants (such as conifers and tobacco) and are also found in insects such as ants and termites, where they show pheromone-like actions and are part of their chemical defense arsenals. By far the richest source of cembranoids to date has been marine invertebrates. Most marine cembranoids are from soft corals (gorgonians), where they comprise about 25 percent of their secondary metabolites. Interestingly, cembranoid-like molecules are not limited to plants or invertebrates, having also been found in the paracloacal glands of the male Chinese alligator (*Alligator sinensis*). As of 2020, some five hundred naturally occurring cembranoids have been described.

Cembranoids have an interesting story—while they were discovered in corals back in the 1960s, to date we still do not know exactly what function they have in the organisms that produce them. But despite the relatively few biological effects that have been documented in the scientific literature, cembranoids display a lot of promise in the area of biomedical research. Some of these compounds have anti-inflammatory or antimicrobial properties, some have anti-tumoral activity, and yet others are neuroprotective in experimental models of stroke, Parkinson's disease, and pesticide neurotoxicity.

Cembranoids are a group of compounds very close to my heart.*[4]

One thing is for sure, no organism evolves a chemical just to give us scientists the excitement of discovering it. As we've said again and again, if a molecule has been conserved, meaning that organisms have continued making it over their evolutionary history (frequently counted in millions of years), it follows that it *must* have a natural function, likely upon predators or prey—and just like organisms in terrestrial environments, organisms in marine environments are bound to employ strategies to make use of (or to defend against) these chemicals.

We know a bit more when it comes to reptiles. Besides the afore-mentioned example of the Indian garden lizard, we have experimental studies suggesting that if other reptiles did take drugs, they'd likely be affected much as we are. When red-tailed boas (*Boa constrictor*) are exposed to benzodiazepines (the class of drugs that includes Valium), they react in exactly the same way as mammals do.[5] And opioids appear to relieve pain in bearded dragons (*Pogona vitticeps*) and corn snakes (*Elaphe guttata*).[6] This may not sound very surprising to you, and indeed it isn't, terribly. Based on the close structural similarities in the nervous systems of vertebrates, it stands to reason that reptiles and mammals would display similar neurochemical and neuropharmacological profiles—and there is no doubt in my mind that we will find examples of shared neuropharmacological features between mammals and fish and amphibians as well. Because the animals covered in this chapter are

---

* Cembranoids were the topic of my master's thesis in Puerto Rico, and in more than one way "kick-started" my scientific career; they are still part of my research program to this day. A thorough review of what they are and do is beyond the scope of this book, but if you are curious about them, and about the story of how (guided by a few wonderful mentors) I found myself working with these compounds, I have placed a few general references in the endnotes.

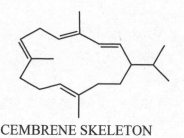

CEMBRENE SKELETON

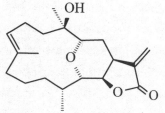

EUNIOLIDE                    EUNICIN

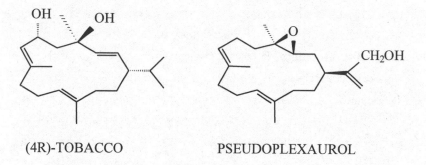

(4R)-TOBACCO            PSEUDOPLEXAUROL

*Figure 6.1. Cembranoid examples. Drawn by the author.*

all vertebrates, all have some basic neurological similarities that make it seem like not much of a stretch to say that, for instance, a drug we use as a painkiller will likewise relieve their pain. Then again, while it may be easier to see yourself in your dog than in the ant crawling across your kitchen floor, there is no doubt at all that even dogs are very different from us, and—as we have discussed at length—there is only so much we can know about how they experience the world. Especially when that experience is colored by a psychoactive substance.

# STOP THE PRESSES!
## ALLIGATORS ON METH?

Environmental toxicology is a flourishing science for all the wrong reasons. I do not mean this as a slight to the science of toxicology; rather, this is a field that exists mostly because of what we humans do to the planet we live on. Our species releases thousands of different chemicals into the environment, and it is the job of environmental toxicology to monitor and study the effects of these compounds on living systems. After all, the proverbial circle of life will ensure that these chemicals come back to us in one way or another.[7]

A subset of this growing discipline involves the effects of prescribed or abused drugs. The most common way these substances get into the environment is via our waste products, which go down the sewers after they make their way through our natural plumbing. Even though the amounts of these chemicals released by a single human are small, there can be thousands of humans in a community, and these small amounts add up. There are also times when much larger amounts are dumped down the pipes, usually in an attempt to get rid of illegal drugs in a hurry. This is not a new practice—people have been doing this as long as we've had access to toilets and abused drugs*—but it is only recently that we have begun to consider what happens to these drugs once they enter the ecosystem. After all, these substances or their metabolites affect multiple species in the environment. This is a pressing problem to be sure, but sometimes, concern about what might get into the water causes . . . overreactions.

---

* Never in a million years did I expect to write a phrase such as this.

In July 2019, law enforcement officers in Loretto, Tennessee, went to arrest a suspect and, upon arriving at the suspect's house, found said suspect trying (unsuccessfully) to flush a significant quantity of methamphetamine (Figure 6.2) down the toilet. After an arrest was made and the contraband confiscated, the chief of police issued a public service statement, urging people not to dispose of drugs in this way because of the possible effects on the environment. The statement ended with a warning that, if the drugs made it far enough, "we could create meth-gators." [8]

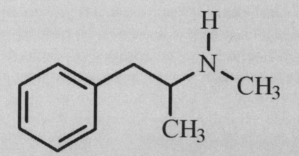

*Figure 6.2. Representative methamphetamine molecule. Drawn by the author.*

Now, there is absolutely no evidence of improper drug disposal creating amped-up alligators on meth, but it is possible at least in principle—and the public service message about drugs and the environment is a sound one. Alligators are scary enough without chemical help; the thought of an angry, meth-fueled alligator gives me the chills.

## BLITZED BIRDS

It is not uncommon for birds to collide with large objects when flying at night. However, it *is* odd to witness a bird colliding with an unambiguously big structure (like a building) when flying during the day. When this happens, it is usually an indication that something is wrong with the aforementioned bird's sensory system. And in more than one case, alcohol has been blamed as the culprit.

The cedar waxwing (*Bombycilla cedrorum*) is a beautiful social bird that displays a characteristic red color on the tips of its wings.[9] In 2012, Dr. Hailu Kinde and his collaborators at the California Animal Health and Food Safety Laboratory System (CAHFS) in San Bernardino reported a detailed study of several flocks of cedar waxwings that died under unusual circumstances—after colliding with objects like plexiglass, fences, and windows while flying during the day.[10]

Upon examination, necropsies* indicated that most of the waxwings had consumed significant amounts of overripe fruit from a Brazilian pepper tree (*Schinus terebinthifolius*). Analysis of the bodies showed extensive hemorrhage in their muscle tissue, oral cavities, and upper digestive systems; most of the examined birds also showed liver rupture. In the paper, the cause of death was ruled as "trauma that resulted from colliding with hard objects when flying under the influence of ethanol." The examiners noted that some birds had alcohol concentrations close to 0.1 percent in their livers, with somewhat lesser amounts in other tissues. (The legal limit for driving according to US federal law is a blood-alcohol concentration of .08 percent.) The same report also summarized six other cedar waxwing "mass deaths" (defined in various ways, but usually as meaning more than five and up to fifty birds) that occurred in the same geographical area over a period of fourteen months between 2005 and 2007. In four of

---

* In general, the term is *autopsy* when performed on a human and *necropsy* when performed on an animal.

these six cases, analysis found that the birds had eaten pepper tree fruit—and in all six cases, the birds presented with ruptured livers. Kinde's team argued that there was "strong circumstantial evidence" for ethanol intoxication, and it is hard not to agree.

This was not the first time that birds in general or waxwings in particular had been suspected of "flying under the influence." Between the 1980s and early 2000s, several studies tackled this idea, providing evidence that alcohol contributes to flying accidents in various bird species.[11] However, I suspect that one of the earliest reports of drunken birds dates from much earlier, though it is unrecognized as such. In 1936, Dr. Mary Louise Fossler of the University of Southern California studied the case of possible mass death involving a flock of cedar waxwings that lost a significant number of birds in the vicinity of a few date palm trees (*Phoenix canariensis*). This particular flock (about five hundred birds) fed on the fruit of these trees after a period of hard rain. Based on Dr. Fossler's description, the dates had been exposed to water for some time; she described them as "water-soaked."[12] The birds themselves were described as "feeding greedily" on the fruit and, after a while, some (but not all) of the birds began to show signs of intoxication, such as lack of coordination, as well as (strangely) suffocation-like behaviors. Then a number of birds simply and unceremoniously dropped dead (not because of any "colliding accident"), dutifully followed by many of their flockmates.

The next day, Dr. Fossler performed necropsies on some of the specimens, recording observations that closely parallel those reported by Dr. Kinde's group almost seventy years later. There was blood accumulation in the major organ systems. Moreover, the birds' livers were especially affected in this way, with signs suggestive of liver rupture. Dr. Fossler hypothesized that the cause of these findings was poisoning due to cyanide-containing compounds in the fruit's seeds. Now, it is true that the seeds of certain plants do contain cyanide derivatives (albeit in small quantities), and it is therefore conceivable that a small bird might get cyanide poisoning upon ingesting these. Alas, I find no reports of this particular species of date plant producing any cyanide-like compounds (although they are known to undergo alcoholic fermentation), deepening this mystery. And still (I am thrilled to have this opportunity to write the following literary phrase) the plot thickens. Sometimes, chemicals like the aforementioned cyanogenic compounds act as deterrents against birds and other grazing animals. In the case of waxwings and similar birds, certain alkaloids are indeed very efficient repellents,[13] and yet cyanogenic compounds were specifically found *not* to act as deterrents (likely meaning they are less toxic) to our friends the cedar waxwings.[14] Taking all these facts together, my unapologetic guess is that alcohol intoxication (or outright alcohol toxicity), rather than cyanogenic poisoning, might well be at the root of this 1936 case of mass death among cedar waxwings. It would be interesting to see if any of Dr. Fossler's bird specimens have been preserved, just waiting to be analyzed with modern biochemical techniques . . .

The above stories underscore both the importance of moderation when consuming a toxic substance like alcohol and the risks of operating a motor system while drunk, but drinking is not always a death sentence. Let's lighten the mood a bit with a look at another of alcohol's effects on our feathered friends—one with which we may be quite familiar.

## Drinking Song(bird)s

I need hardly tell you that music is an integral part of every human culture that has ever graced this planet. Music can express feelings of love and hate, sorrow and happiness, patriotism and nostalgia, among many others. I also need hardly tell you that feelings, whether in private experience or public ritual expression, are a frequent impetus for alcohol consumption in humans. It is no surprise, then, that music and drinking coalesce in many human populations when trying to express the myriad emotions that make up the human condition. In fact, there is a whole genre devoted to the integration of singing and alcohol: *drinking songs*.

Drinking songs are frequently sung in a ceremonial or recreational setting, oftentimes by a group of friends (or even very recent acquaintances, say at a bar on New Year's Eve) as a way to bond over some of those aforementioned feelings. As these songs are not usually performed for a paying audience, zeal is prized over skill, and in general, technique is the least important component of drinking songs (strict adherence to the lyrics is often optional as well). This is a good thing, as excessive alcohol consumption impairs motor functions, and (as anyone who has been in a bar on New Year's Eve can attest) singing undeniably belongs in this category. In other words: *drinking song singers sound drunk.* (Try saying that three times fast! I'll bet you can't, especially if you've been drinking.)

Certain bird species, notably—and rather obviously—songbirds, are very proficient at singing, and so you might be interested to hear that when under the influence, they sing drunkenly too! A paper by Dr. Christopher R. Olson and his collaborators at Oregon Health & Science University, with the right-on-point title "Drinking Songs: Alcohol Effects on Learned Song of Zebra Finches," detailed the effect of ethanol on the quality and structure of the songs sung by zebra finches (*Taenopygia guttata*).[15] The zebra finch is not just any songbird; it is considered a near-perfect animal model in which to study the motor integration of complex tasks, like singing, and its associated neurobiology. Finding an animal model for "vocal learning"—which is vital to human language development—is difficult; many mammals with whom we share other cognitive similarities are, well, rather taciturn. Zebra finches are not only one of the very few species that engage in vocal learning, but the way they learn and produce their song is surprisingly like the way humans learn and produce speech.[16]

So, familiar with the effects of alcohol on human speech, Dr. Olson and his friends decided to get these birds drunk. They were surprised to find they had little trouble convincing the finches to drink (in fact, the birds given access to alcohol-spiked juice increased their fluid consumption). And once they'd drunk . . . they sang sounding drunk. Their singing was a bit sloppier, a bit less organized. It was also a bit quieter (a sharp contrast to my experience of drunken humans), and the more alcohol consumed by a particular bird, the more maudlin its song sounded. One of the best-studied characteristics of the songs of these birds is their complexity, with distinct "syllables" easily identified; interestingly, alcohol consumption affected different syllables unequally. This is much like when humans singing (or speaking) drunk pronounce certain words without any problem, but find the pronunciation of other words more challenging. For example, occasionally, when a human-led drinking song is in full swing, the "uttering" degrades to "muttering." Ditto for songbirds. Also curious was the fact that the birds had alcohol concentrations around the legal limit for humans, but even when their singing was affected, their flying, perching, and similar motor skills did not seem to suffer, suggesting that other neural areas were involved.

To date, I know of no other studies on the effects of alcohol on songbirds, but I believe that more research into birds and alcohol is forthcoming. Since I am blessed with an overactive imagination, I cannot help but think what a difficult situation it would be for a scientist (or anyone, for that matter) to handle a drunk and possibly belligerent ostrich.* One thing is certain, the Olson paper offers "proof of principle," showing without any doubt that birds can and do get drunk and live to tell the tale (if they remember it, that is).

---

* There are videos online purporting to show drunken ostriches, but I would take those with a big grain of salt—maybe the same sort you would use to rim a margarita . . .

## HAMMERED BATS

Next up, an altogether different example of a flying organism on booze. We met Dr. Robert Dudley earlier in the book when describing his drunken monkey hypothesis; in 2004 he and his collaborators at Ben-Gurion University in Israel (led by Dr. Francisco Sánchez) tested that hypothesis using the Egyptian fruit bat (*Rousettus aegyptiacus*).[17] Specifically, they tested whether frugivory (fruit eating) in this bat species is influenced by ethanol—that is, whether the alcohol content of ripening fruit served as an olfactory cue and attracted the bats.* The researchers took samples of the fruits of four different plants (sycamore fig, Qordi, date palm, and jujube tree) and measured the percentage of ethanol by volume in their ripened fruit, which ranged between 0.1 and 0.7 percent. Since *R. aegyptiacus* actively feeds on those four fruits, among others, the researchers assumed they preferred that range of ethanol concentration. Working from this assumption, they tested whether ethanol dissolved in water or fruit juice attracted the attention of captive bats. Alas, they found quite the opposite; not only were the bats indifferent to ethanol dissolved in water or juice up to a concentration of 1 percent, but they were actively repelled by alcohol-laced liquids with an ethanol concentration of 2 percent.

In light of these results, Dr. Sánchez (the lead author of the aforementioned paper) became curious about whether this bat species actively avoided ethanol-rich food in order to avoid being impaired, and conducted a study to ascertain whether the flight abilities of these bats were affected by alcohol.[18] The researchers predicted that ethanol would affect two factors in these bats: the time it took them to cover a given distance and their echolocation behavior (the means by which they navigate). To get the bats drunk, they fed them a tasty concoction made from a commercially available infant formula with

---

* Part of the idea of these experiments was the notion that bats help disperse the seeds of the fruits they eat; this could suggest an evolutionary relationship between bats and these fruit-bearing plants.

or without 1 percent alcohol by volume (obviously, they added the ethanol themselves aftermarket). They then measured the bats' speed and recorded their echolocation sounds both when fed regular food and when fed that containing alcohol. Not surprisingly, they found that tipsy bats flew more slowly than those who were sober, and their echolocation patterns changed as well—thus, drunkenness in nature might well spell doom for these bats, as their ability to locate and catch food would not be performing optimally.

Interestingly, the results described above seemed to apply only to this one species of Old World bat. Dr. Dara Orbach of the University of Western Ontario in Canada and her collaborators reported a series of experiments testing the abilities of six different species of New World bats to fly and echolocate when tipsy.[19] To make a long story short, no bat in these experiments became impaired either in their flight capability or in echolocation, even upon consuming food laced with ethanol up to a concentration of 1.5 percent. Now, consider, please, the complex computations that any echolocating bat must engage in when determining where to fly and how fast (avoiding obstacles along the way), along with the highly coordinated muscle activity that is essential to flight. This remarkable ability of the bats of the Americas is surely due to a relative degree of resistance to ethanol, which in turn must be related to increased metabolism or optimized detoxifying pathways. These results represent an interesting "intercontinental" difference between the relative capacities of bats to withstand ethanol, and could lead to fruitful insights on the separate evolutionary paths of these animals.

Speaking of paths, let's return to the ground, and the animals that live there. And let's think bigger—in fact, let's start our exploration of the grounded at the top, with the largest living terrestrial animal.

## ELEPHANTS UNDER THE INFLUENCE

Alcohol has been part of the lore of elephant behavior for hundreds of years. There are well-documented instances of elephants getting ahold of locally produced alcoholic drinks, getting unequivocally drunk as a result, and getting rowdy (or worse) because of it—elephants have even killed people in this drunken state. But while it is clear that when they happen upon a chance to consume alcohol, they take it, what is less clear is whether they seek it out on purpose. Many people think they do, and a leading hypothesis is that elephants are quite familiar with alcohol and get drunk without human help by ingesting fermenting fruit. There are informal descriptions of Asian elephants seeking out the fruit of the durian tree (usually *Durio zibethinus*), especially when fermented,* as well as tales of African elephants consuming the fermented fruit of the marula tree (*Sclerocarya birrea*) with the purpose of getting drunk. To date, I have been unable to find published scientific reports supporting either case, although there seems to be some circumstantial evidence in the case of African elephants and marula—in fact, this story offers a perfect example of the surprises nature so often has in store for science.

One of the earliest reports of this behavior comes from an 1800s account by the French naturalist Adulphe Delegorgue, who collected and transcribed stories from his Zulu guides telling of male elephants who became aggressive after gorging on the marula fruit. Delegorgue wrote: "The elephant has in common with man a predilection for a gentle warming of the brain induced by fruit which has been fermented by the action of the sun."[20]

As we have said many times, it is tempting to ascribe human motives to a nonhuman animal, but there is no indication that Delegorgue had any evidence beyond the reports of his guides—and,

---

* The durian fruit is infamous for smelling really bad; can you imagine how bad it smells when in addition to its natural stench it is overripe and fermented?

when it comes to elephants getting drunk on fermented marula fruit, there is some scientific evidence arguing against this notion.* In 2006, Dr. Steve Morris (from the University of Bristol) published a paper with several collaborators in which they evaluated the claim, looking at whether African elephants seek out marula tree fruit for food, whether they prefer it when fermented, and whether they might become intoxicated as a result.[21] Here are the facts: There is no doubt that marula fruit (which is similar to mangoes) is a preferred source of food for elephants. However, they tend to take it directly from the tree, where it is unlikely to be fermented. There are even reliable reports of elephants shaking marula trees to make the fruit fall, proceeding to collect and eat it afterward. On the other hand, yes, sometimes elephants do feed on decaying and presumably fermenting fruit. Despite this, hard data pointing at a direct relationship between fermented marula consumption and elephant drunkenness is sorely lacking. Dr. Morris's paper analyzes how much fermented fruit a fully grown elephant would have to eat to become inebriated, as well as whether non-ripe fruit could ripen and ferment during its long voyage through an elephant's gastrointestinal tract (which usually takes between twelve and forty-six hours). The analysis assumed that alcohol metabolism in elephants and the amount of ethanol required for inebriation would be similar to those factors in humans, and scaled to account for our size difference—an adult male elephant can weigh up to thirteen thousand pounds, while a typical adult male human might weigh about two hundred pounds. Their calculations suggest that three conditions would have to occur simultaneously in order to get a fully grown male elephant drunk exclusively from eating fermented marula fruit: (1) the elephant would have to eat unrealistic amounts of said fruit (and nothing else); (2) marula fruit would have to produce even more unrealistic amounts of alcohol; and (3) any fermentation in the elephant's gut

---

* I know of no formal scientific reports about Asian elephants and durian fruit.

would have to exclusively yield ethanol (and none of the other usual substances, like methane). Even if we imagine that an elephant ate nothing but this fruit and that it has an off-the-charts fermentation capability, this paper suggests that natural inebriation as a result of eating marula fruit is at best unlikely, unless ethanol metabolism in elephants is fundamentally different than ethanol metabolism in other mammals, including humans.

However—plot twist!—it turns out that this might just be the case. Dr. Mareike Janiak and colleagues at the University of Calgary performed a genetic analysis of eighty-five mammal species, looking at certain genes and gene variations involved in the metabolization of ethanol, specifically those that encode our good old friend the enzyme ADH. One variation they studied was the same one we discussed in chapter one, which is present in humans and our ape ancestors, as well as in some members of an entirely different evolutionary branch, like the slow loris and the aye-aye. As you may recall, this variation increases the efficiency of ethanol metabolism by some 40 percent. Well, not only did elephants not have this variation, they lacked the gene in question entirely! As Dr. Janiak notes in an article discussing these findings in the context of the elephant/marula controversy, "[I]t's possible that elephants have another way of breaking down ethanol. But it's very unlikely that the efficiency with which they can do this is comparable to that of humans. Simply scaling up for body size does not accurately predict whether elephants can become intoxicated from eating old marula fruit."[22] As it happens, several other mammal species also lack this gene. In the 2020 paper publishing their results, the authors note in an ending comment that "humans have a tendency to anthropomorphize animal behaviours. However, it is similarly a mistake to assume that animals share our metabolic and sensory adaptations or limitations."[23]

For now, the mystery of the (allegedly) marula-mad elephants remains unsolved. However, we have already seen that when these animals get their hands (er, trunks) on *human*-made alcohol, they are perfectly capable of consuming enough to get drunk and act accordingly. I do not expect you will be surprised to hear that some brave scientists eventually took it upon themselves to get elephants drunk *on purpose*. Moreover, we have talked about one of these fearless scholars earlier in this book—Dr. Ronald K. Siegel, who we met in chapter one.

In a 1984 paper, Dr. Siegel and his colleague Dr. Mark Brodie reported on the self-administration of ethanol by elephants.[24] More precisely, this was scientist-facilitated self-administration. Siegel and Brodie's subjects were three Asian elephants (*Elephas maximus*) and seven African elephants (*Loxodonta africana*), all born in the United States and raised in a game park in California. They (the researchers, not the elephants) provided the refreshments: a water solution containing 7 percent ethanol, unflavored, and delivered via a big metal drum in the back of a Jeep. The elephants were allowed to drink at their leisure, and some were evidently very enthusiastic about it, since at least one of the elephants single-handedly (-trunkedly?) drank about 75 liters of the offered beverage. Siegel and Brodie then observed the behaviors displayed by the elephants, analyzing them based on data from the published literature on elephant behavior in general. The parameters they measured and scored included aggression, bathing, exploration, feeding, rocking, vocalization, ear flapping, and "inappropriate behaviors," which the authors described as "wrapping their trunks about themselves and both leaning and stationary postures with closed eyes." (I know, it doesn't sound so bad to me either.) After imbibing, the elephants showed a wide variety of effects: some became mellow, some became aggressive, some became uncoordinated to the point of not being able to remain upright, but all of them displayed behavior not usually expressed when "sober." In the authors' words, alcohol seemed to "bring out the individual personality of each animal." Interestingly, the more stressed the elephants were (as inferred by an increase in vocalizations and aggressive behaviors), the more they drank, though why that was so is of course unclear—as far as I know, nobody asked. To my knowledge, no similar studies on ethanol consumption in elephants have been done since.

## Of Elephants and Vulcans

The fictional universe of *Star Trek* has influenced our real world for fifty-plus years. There have been other fictional universes: some of them darker, some of them more idealistic or, on the other hand, more realistic; some of them funnier and more fantastic. Although I am a fan of many of these other universes, the universe of *Star Trek* has a special place in my heart, and I know that I am not alone in this.

One of the most beloved characters in this universe is Mr. Spock, the science officer of the USS *Enterprise*. As the story goes, Mr. Spock was the offspring of a human mother and a Vulcan father. In *Star Trek* lore, the Vulcans are an alien species that essentially had to embrace logic or die—they were a rather violent race, a trait that proved especially dangerous because they were also highly intelligent, technically adept, and physically powerful thanks to evolving on a rather inhospitable planet. Thus, on the verge of killing one another off, from a certain point in their history onward the Vulcan society adhered to a strict code of logic, suppressing their emotions in order to preserve their species. Their logic-led existence had one lone exception: a recurring period in the life of every Vulcan when, let's just say, *biology rules*. More specifically, according to *Star Trek* canon, approximately every seven years or so Vulcans undergo *pon farr*, a period of physiological instability that results in a mental imbalance directly related to reproductive matters.[25] During pon farr, Vulcans, male and female alike, experience an extreme and hard-to-control—you might say irrational—urge to mate (it is a tad more complicated than that, but this suffices for the purposes of our story).[26]

Don't worry, I haven't forgotten we are supposed to be talking about elephants on drugs. It just so happens that male elephants experience a physiological phenomenon that in my mind is reminiscent of the fictional pon farr. This phenomenon is called *musth*—it occurs about once a year in male elephants, and it is characterized by a number of behavioral changes including extreme aggression, enhanced sexual drive, and general irritability.* Elephants during musth produce a variety of hormones and unique secretions, and in fact may display testosterone levels about sixty times their normal values. There is now research indicating that a surge in male hormones occurs *before* the behavioral manifestations of musth, strongly suggesting a causal relationship between this surge and musth's onset.[27] Additionally, musth seems to correlate with changes in thyroid hormones, as well as abnormalities in sugar metabolism. (No wonder elephants in musth are irritable—*I* would be!)

When in musth, a male elephant will frequently fight other males to gain access to (and the favor of) the fairer sex. As you might imagine—with all those surging hormones, active belligerence, etc.—musth is a metabolically expensive state for elephants, and they cannot sustain this energy-taxing condition for very long. Some scientists have even speculated that female elephants might be more receptive to males in musth not only because of their displays of dominance, but also because a male able to sustain this state must be healthy. In other words, musth could be a strategy that male elephants use to advertise their physical fitness to sexually receptive females.† If the big fella is unlucky in matters of love, however, he becomes "frustrated" and even more volatile. No one will argue with the notion that a bull elephant in musth is one of the most dangerous terrestrial animals on this planet. As we saw above, male elephants (African or Asian) can

---

* "Musth" is a Sanskrit word meaning "intoxicated," although elephants display a behavior in this state that is clearly more than mere drunkenness, a fact that plays an important role in this story. Musth occurs in both Asian and African elephants, two species that diverged about five million years ago, about two million years after the divergence event that separated our line from that of the chimpanzees.

† As a fellow male, I admit that many of us guys do like to show off, *especially* when trying to impress a prospective mate. We just use slightly different strategies, that's all.

tip the scales at 13,000 pounds (more than six tons); in comparison, an SUV on the larger size range weighs about five thousand pounds (two-and-a-half tons). Elephants in musth may throw elephant-sized tantrums, wreaking havoc on anything and everything in their paths, with no apparent thought process or reason guiding their actions. In fact, some scholars have compared musth with psychotic states.

Just as animal models are invaluable to the study of normal physiological processes, the "Holy Grail" of most areas of medical research is the identification of animal models mimicking particular pathological states that enable scientists to perform research applicable to human health. Probably in no medical specialty is this goal more challenging than in psychiatry, and it is not hard to understand why. Psychiatric conditions, while undeniably neurological in nature, nonetheless often involve subjective phenomena and can be very hard to diagnose in humans, let alone in experimental animals. Take, for example, a group of symptoms collectively called *psychosis*. The very definition of psychosis, "conditions that affect the mind, where there has been some loss of contact with reality,"[28] is a clear indication of how subjective psychotic disorders—disorders, like schizophrenia, that involve psychosis—can be. Psychiatrists must diagnose such conditions largely through behavioral observation and (to some extent) self-reported experience,* since there are no clear-cut physiological markers of such states. In other words, there is no blood test that would help diagnose psychosis, as a physician might use to diagnose diabetes or high cholesterol. In large part because of the current lack of biochemical correlates for this set of mental conditions, the development of useful medications to treat them presents a distinct challenge, especially considering that—for reasons like those described when discussing animal "hallucination"—there is no animal model useful to understanding human psychosis. This is where

---

* In the words of a famous (and fictional) medical genius: "Everybody lies."

the similarities between musth and the behaviors associated with certain well-described human states of mind come into play.

In 1962, two psychiatrists at the University of Oklahoma School
of Medicine, Dr. Louis J. West and Dr. Chester M. Pierce, had what
at the time seemed like a very good idea.* They'd put together a few
otherwise disparate pieces of information to construct a scientific
hypothesis. First, they knew that musth in elephants expressed itself
as behaviors very much like madness, or in more technical terms, psychosis. As psychiatrists, they were aware of several drugs that induced
similar states in humans, most notably LSD. Thus, they hypothesized
that in a male elephant, LSD might trigger musth.† To test their
hypothesis, they teamed up with the director of the Lincoln Park
Zoo in Oklahoma City, Warren D. Thomas, who "volunteered" one
of the zoo's inhabitants as a subject: Tusko, a fourteen-year-old male
Asian elephant who weighed close to 7,000 pounds (about 3,200 kg).
One has to admit that these were adventurous people—after all, their
plan was to intentionally induce madness in an animal heavier than
an SUV. But West and Pierce did not go into the experiment blindly.
They carefully considered what dosage to give to Tusko, making an
educated guess based on the dosages of LSD that people tended to use
recreationally, as well as on the amounts of the drug that had proved
lethal to other experimental animals like cats and rats. This being said,
based on the zookeeper's experience giving drugs to elephants in the

---

* This phrase brings back fond memories of graduate school. My PhD adviser
at Cornell, the late professor George P. Hess, would regularly go around the
lab to ask us about our experiments. You would better have something to show
him when he asked, "So how's the data?" From time to time, though, he would
go straight to any of us and state: "I have a very good idea," meaning an idea for
an experiment, which to us would sometimes mean a significant sidetrack from
our ongoing experiments. (Almost invariably, though, it really *was* a good idea.)
† This was the "formal" justification for their experiment. However, there was
surely a strong element of simple curiosity involved. As the author Alex Boese
says on page 113 of his 2007—and very aptly titled—book, *Elephants on Acid
and Other Bizarre Experiments*: "After all, 'what would an elephant on acid do?'
It's hard not to be curious."

wild, they favored erring on the higher rather than the lower side, and eventually decided to administer close to 300 milligrams—about 0.1 mg/kg in pharmacological parlance. A weight-based dose of 0.1 mg/kg is many times higher than that usually taken by humans in a typical recreational session (these doses are measured in micrograms, and might work out to something like .002 mg/kg at most), but below the lethality threshold observed in other experimental animals. The scientists essentially extrapolated a "safe" dose based on the "safe" dosage in other organisms and administered it proportionally by weight. However, weight is not the only factor involved in determining drug dosage, and a recent analysis based on a more nuanced understanding of scaling calculated that the "safe dosage" for Tusko would have been closer to 32 mg (about .01 mg/kg), an almost tenfold difference from what the psychiatrists gave the gentle giant.[29]

West, Pierce, and Thomas narrate the story in detail in their paper reporting the experiment,[30] but in a nutshell, this is what happened: At 8:00 a.m. on the morning of August 3, 1962, they shot Tusko with a dart containing 297 mg of LSD. The events that followed Tusko's exposure to LSD ensued in rapid succession, and everything was over in less than two hours. Briefly, Tusko became restless for a few minutes, and then markedly uncoordinated. Soon after becoming uncoordinated, Tusko lost the ability to stand upright.* He fell (hard) on his side, and (among other things) began to experience seizures and difficulty breathing. These symptoms, particularly the seizures, triggered a frantic response from West and Pierce, who proceeded to inject Tusko with massive amounts of an antipsychotic medication, followed by a similarly large dose of barbiturates.

Despite these efforts (perhaps even because of them), Tusko died at about 9:40 a.m.

---

* Incidentally, Tusko had a mate, Judy, a female Asian elephant who apparently came to help and comfort Tusko when he became visibly affected by the drug. (By the way, what were they thinking? They left Judy in the same enclosure as Tusko, fully expecting him to go psychotic!)

Tusko's death was a mystery at the time, and in fact still remains so. While the dose they gave Tusko was indeed high, other animals had been given (and handily survived) similar doses by weight. Although West and Pierce could not rule out a special sensitivity of elephants toward LSD (and in fact this was their main conclusion), this is unlikely based on the observed effects of this drug in a variety of mammals (humans included). On the other hand, LSD is a powerful drug in the sense that only minute amounts can trigger marked behavioral effects, at least in mammals, and drugs can and do kill creatures through mechanisms other than outright toxicity. It is clear that West, Pierce, and Thomas had high hopes for their experiment and planned to do more research on elephants as a possible animal model for mental diseases. In their 1962 paper on Tusko, they refer to either a paper or a book titled "Of Elephants and Psychiatry," listing it as "in preparation." It was never published.* By the way, West, Pierce, and Thomas did not observe any behaviors in Tusko that were even remotely reminiscent of musth. Sadly, it seems that he died in vain.

In the end, the only way to know how LSD affects elephants (Was it lethal? Did it cause musth after all?) was to repeat the experiment, which Dr. Ronald K. Siegel did in the 1980s. However, Dr. Siegel made a few procedural adjustments. His experiment used two elephants, one male and one female, weighing about 4,400 pounds (2,000 kg, roughly two-thirds of what Tusko weighed) and 3,300 pounds (1,500 kg), respectively. The LSD was given in the elephants' drinking water rather than injected, and, most importantly, each elephant got two different doses (administered in separate sessions): one "low dose" of about 0.003 mg/kg, and one "high dose" of 0.1 mg/kg (the same as Tusko).

---

* Just over two decades later, a paper critical of the Tusko experiment was published with precisely this title: Jentzsch (1983) "Of Elephants and Psychiatry," *Freedom*, No. 58, 6–7.

Seigel published his results in 1984 in the *Bulletin of the Psycho-nomic Society*.* He stated his rationale for the experiment as follows:

> *Because the original study has continued to receive attention and criti-cism (e.g., Jentzsch, 1983), the following study was conducted to correct procedural problems and reexamine the relationship between LSD and musth behavior in the elephant.*

I do not doubt that Dr. Siegel's main motivation was scientific research, but I bet that he was also curious to see what would happen if he gave the "correct" amount of LSD to an elephant.

On the low dosage, both elephants displayed what Seigel refers to as "dramatic changes in behavior" within twenty minutes. In the case of the female, these effects are described as "a small increase in rock/sway time and slightly increased ear flapping and exploration" along with an increase in "inappropriate behavior" (that "leaning with closed eyes" again!) and an uncoordinated gait. (Scientists sometimes have a rather low threshold for what they consider "dramatic," it is true.) The male's behavior on the low dose was much the same, but more pronounced, and with the addition of a few "aggressive displays." These "aggressive displays" seemed to be mostly for show, meaning they were confined primarily to behaviors like vocalization—he never attacked the female elephant, and it was nothing like musth.

On the high dose—the same weight-adjusted dosage as that given to Tusko—the male once again made an "aggressive display" (trum-peting and snorting while charging the observer), then moved on to leaning with closed eyes and seeming uncoordinated, occasionally taking a break for another aggressive display or a spot of dust bath-ing. The female's response to the high dosage was more eventful, with

---

* *Psychonomics* is the general area of psychology whose explicit goal is to "figure out the laws and principles that control the phenomenon of mind." You gotta hand it to these scholars; they didn't have an ounce of pessimism in them.

an initial aggressive display of trumpeting followed by an increasing lack of coordination that progressed until she fell over. She stayed down for sixty minutes, breathing shallowly and having occasional tremors, and for a while, it must have seemed worryingly close to a replay of Tusko's fate, but when the handlers nudged her (the fact that they waited an hour to do so suggests she can't have looked that bad off, I suppose?) the female elephant "arose slowly and eventually regained an upright posture." The effects of LSD disappeared within twenty-four hours in both elephant subjects. Dr. Siegel concluded that LSD administration to elephants "provides an unsatisfactory model for the natural aggression and behavioral disorders associated with musth." And it seems that, while the most likely explanation remains an LSD overdose, we will likely never know for certain what *really* killed Tusko.

I have no desire to gloss over the darker aspects inherent in the history of this book's topic, and feel that Tusko's story is an important one that merits its inclusion (at some length). We can take comfort in knowing that this experiment would never be approved today, due to the happy fact that there are now many more regulations on the administration of experimental drugs to vertebrate animals. Our next story is a far less sobering one, involving animals taking drugs of their own volition—in this case, some of them creatures I know personally.

## CHILLING (OR NOT) WITH SOME CATS

A relatively common psychoactive substance that affects animals—and one with which you are likely to be familiar—is *catnip*. *Nepeta cataria* is a plant found the world over, and it was used in traditional medicine as an antipyretic (anti-fever) remedy in Europe for millennia.[31] The main active component of catnip is nepetalactone, which is currently being investigated as a mosquito repellent,[32] as an antimicrobial,[33] and as an analgesic,[34] among other applications of interest to humans. However, the best-known "targets" of catnip in general and nepetalactone in particular are cats, big and small. Despite the afore-mentioned medicinal effects of nepetalactone, as far as we know, cats actively pursue and consume catnip for pure pleasure.

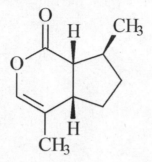

*Figure 6.3. Nepetalactone. Drawn by the author.*

I have two grandcats; their loving human is my daughter, Giselle Vanessa (who is very much loved herself). Her first cat was Eluney,* a lovely, affectionate white-and-black male who nonetheless looks permanently pissed-off thanks to a natural expression that gives the dearly departed Grumpy Cat a run for his money. A few months after Eluney, Vanessa adopted Elara,† a sweet-looking calico beauty; alas, she is much, and I mean *much*, grumpier than Eluney. Her vet (Elara's, not Vanessa's) told Vanessa that the demeanor of calico females tends to be on the grumpy side in general, but as a dog person myself, I wouldn't know.

---

* His name means "Gift from Heaven" in the native language of Taíno Indians from Puerto Rico and the Caribbean.
† Elara is a character in Greek mythology and the name of one of Jupiter's moons.

*Figure 6.4. My grandcats, Eluney (at right) and Elara (at left),
neither under the influence of catnip. Courtesy of she who will
always be my baby girl, my daughter, Giselle V. Pagán.*

The behavior of my grandcats when on catnip is a perfect illustration of one of the principles we mentioned in chapter two, namely
that with any psychoactive drug, the effect on a subject's behavior has a
lot to do with things like the subject's sex and/or their specific genetic
makeup. Upon being exposed to a catnip-containing toy, Eluney displays typical "high as a kite" behaviors; he gazes intently at the ceiling, looks very peaceful, and softly "meows" from time to time as if
immersed in deep philosophical thoughts. In contrast, when Elara gets
ahold of the same toy, she goes absolutely ballistic. She runs around
Vanessa's apartment as if fleeing from an unseen—yet obviously
terrifying—enemy. The influence of the sex of a cat on their response to
catnip is well known,[35] as is the influence of specific genetic variables.

And the behaviors displayed by Eluney and Elara are by no means a complete catalog of the effects that catnip has on house cats. Some other behaviors observed in cats "under the influence" include drooling, raking (kicking movements against the catnip-infused toy), and undulating skin (a wavelike motion of the skin on the cat's back).[36] In both females and males, it is not uncommon to observe spontaneous mating,[37] while another, less active, option is the display of "sphinxlike behaviors," which refers to "time sitting still in a sphinxlike position."* These varied responses help illustrate the wonders of pharmacology. Interestingly, a molecule very similar to nepetalactone functions as a pheromone in aphids.[38]

As you may imagine, house cats are not the only felines sensitive to catnip. Nepetalactone seems to induce psychoactive effects in roughly 60 to 70 percent of a cat population, regardless of species, from house cats to tigers. For obvious reasons, research on the effect of catnip on big cats is relatively scarce, although not inexistent.[39] As expected, along with individual variation, there is a wide range of responses dependent on species as well. In a study on tigers, roughly half of the tested subjects were either indifferent or "responded disapprovingly" to catnip.[40] How does one measure "disapproval" in a tiger, anyway? Does it scowl at you? Does it go "tsk, tsk, tsk" while moving its head from side to side? I will leave this to the scientific literature, thank you very much. Then there is this account of a tiger behaving in what is characterized, deadpan, as an "atypical" manner: "A young tiger, however, took one sniff [of catnip] and leaped several feet into the air, urinating in the process, then fell flat on his back. He scrambled to his feet and dashed head-first into the wall of his cage."[41] The effects of catnip have also been tested on the smallest wild cat in the world, the black-footed cat (*Felis nigripes*), which showed similar (though perhaps less amusing) levels of variation in its responses.[42]

---

* An unrelated fact I am nevertheless compelled to share: the plural of "sphinx" is "sphinges."

A popular hallucination-inducing substance among humans is *ayahuasca*, a brewed tea made from two plants, the yagé vine (*Banisteriopsis caapi*, a source of dimethyltryptamine [DMT], a potent hallucinogen; Figure 6.5), and another plant, usually *Psychotria viridis*, which contains substances that slow the metabolic degradation of DMT. (The combination of both plants enhances the hallucinogenic experience.)[43] One of the most curious cases of big cats on drugs involves accounts of jaguars who consume yagé, the "active" plant used in producing this psychoactive brew, and afterward behave for all intents and purposes as if they were hallucinating.

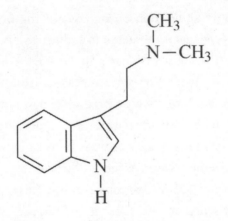

*Figure 6.5. Dimethyltryptamine (DMT). Drawn by the author.*

There is video documentation of jaguars who nibble on yagé vine, and well, they do *look* high. Of course, as discussed many times now, the subjective nature of the experience of hallucination makes it difficult to ascertain the effect of hallucinatory substances on an individual, and this difficulty is even more pronounced when the individual is an animal (please recall the sea slugs in chapter five). There does not seem to be any scientific study into the possible neurochemical effects of yagé consumption on jaguars (who do not complement their yagé consumption with *Psychotria* leaves), and not everyone agrees as to whether the observed effects are truly suggestive of hallucination or if the jaguars are simply behaving as if they were drunk.[44] For now, it is an interesting possibility that warrants more investigation—but not, please, by feeding ayahuasca to your cat.

## CRAZED HORSES

In the mid-eighteenth century, the first written reports of a myste- rious malady affecting horses, cattle, and an assortment of smaller animals began to appear.[45] Within a relatively short time, it became evident that this condition was due to the consumption of twenty-odd species of plants belonging mostly to the *Astragalus* and *Oxytropis* genera,[46] collectively referred to as *locoweeds*. In general, animals avoid consuming these plants due to the bitter-tasting alkaloids that they contain (more on this below), but when food is scarce, as in drought conditions, hungry animals will eat them, and once they do, they act very much as if they were addicted to the stuff. "Loco" is the Spanish word for "mad" or "crazy," which gives us an idea of the apparent effects of these plants—effects collectively known as *locoism*.*

Horses are the best-known sufferers of locoism, and the condition can be dangerous not just to them, but to those nearby as well; imag- ine a two-thousand-pound Clydesdale out of control. Still, probably the best description of the effects of locoweed in horses is one of the earliest ones, an 1874 account by the rancher O. B. Ormsby:[47]

> *I think very few, if any, animals eat loco at first from choice; but as it resists drought until other feed is scarce, they are first starved to it, and*

---

* In an interesting note, it seems that the observation of the effects of locoweed on livestock found its way into the annals of crime: "It is said that when Señora no longer loves her liege lord and fain would rid herself of him, she procures some Herba Loco, prepares a decoction of it and beguiles his innocence into imbibing it, after which he becomes permanently insane or else dies in a short time." From J. Kennedy (1888) "The Loco Weed (Crazy Weed)," Pharmaceutical Record 8:197 (as reported by Bender, 1983).

*after eating it a short time appear to prefer it to anything else. Cows are poisoned by it as well as horses, but it takes more time to affect them. It is also said to poison sheep. As I have seen its action on the horse, the first symptom, apparently, is hallucination. When led or ridden up to some little obstruction, such as a bar or rail lying on the road, he stops short, and if urged, leaps as though it were four feet high. Next, he is seized with fits of mania, in which he is quite uncontrollable and sometimes dangerous. He rears, sometimes even falling backwards, runs or gives several leaps forward, and generally falls. His eyes are rolled upward until only the white can be seen . . . and, as he sees nothing, is apt to leap against a wall or man . . . Anything which excites him appears to induce fits, which, I think, are more apt to occur in crossing water than elsewhere, and the animal sometimes falls so exhausted as to drown in water not over two feet deep. He loses flesh from the first, and sometimes presents the appearance of a walking skeleton. In the next and last state he only goes to and from the loco to water and back, his gait is feeble and uncertain, his eyes are sunken and have a flat, glassy look, and his coat is rough and lusterless. In general, the animal appears to perish from starvation and constant excitement of the nervous system, but sometimes appears to suffer acute pain, causing him to expend his strength in running wildly from place to place, until he falls, and dies within a few minutes.*

The main chemical culprit in locoism (swainsonine, Figure 6.6) was not discovered until 1979, in *Swainsona canescens*, a plant native to Australasia; its discovery in plants native to North America had to wait until 1982.[48] In fact, however, it is not made by these plants themselves, but by symbiotic fungi. As you may imagine, swainsonine and related compounds have a defensive function in locoweed plants.[49] Swainsonine is a pretty molecule (I know, I say this about most molecules), and its mechanism of action is interesting, as it does not seem to target a specific receptor. Rather, it interferes with certain metabolic enzymes by mimicking sugar structures. When the activity of these enzymes is reduced, metabolites accumulate in an abnormal way and wreak havoc, most notably in nerve cells. While there are no evident macroscopic lesions in the nervous systems of the affected animals, microscopic examination reveals neuronal degeneration.

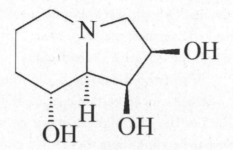

*Figure 6.6. Swainsonine. Drawn by the author.*

The economic impact of locoism was recognized early on, and even today it accounts for agricultural losses measured in hundreds of millions of dollars. The problem is compounded by the fact that there is no known antidote. The best bet for farmers is to try to prevent animals from eating locoweed, and as such, efforts concentrate on eliminating the plant from the environment. Alas, as in the case of most pests, this is easier said than done.

## APOCRYPHAL INTOXICATION: ADDICT BIGHORNS, WEAVING WALLABIES, CAFFEINATED GOATS, AND FLYING REINDEER

This is a somewhat frustrating section for me to write; the stories will also be on the shorter side. Allow me to explain. If you have read this far, you have surely noticed that I take pains to document my assertions, preferably with primary sources such as scientific papers; this is simply my training at work. However, the next few examples of animals on drugs, as interesting as they are, are also notably lacking in scientific documentation in one way or another. Apocryphal or anecdotal accounts of intoxicated animals are everywhere. The fact that they have not made their way into the scientific literature does not mean they are necessarily untrue—as with the example of elephants and marula, we are learning more all the time—but stories are not the same thing as science. It would be impossible to chronicle every apocryphal tale of animal intoxication in this book, just as it is impossible even to chronicle every documented one; these are simply four of my favorites. Let's begin with the tale of the (allegedly) narcotized rams.

Dr. Ronald K. Siegel (who should be an old friend to us by now) described some odd behavior in certain species of sheep thusly: "In the Canadian Rockies the wild bighorn sheep do not roam far for food and stay close to bedding spots they may use for years. Yet they will negotiate narrow ledges, knife edged outcrops, and dangerous talus slides to feed on a mysterious lichen."[50] He went on to tell of these sheep not just relentlessly seeking these lichens but consuming them in an obsessive way. The lichens grow on rocky surfaces, and the sheep are allegedly so intent at scraping the lichen from the rock that they erode their teeth all the way to the gum. Uninterested in other food, the animals eventually die of starvation. Dr. Siegel speculated that the obsessive behavior of these sheep was related to the supposedly narcotic properties of this lichen, as apparently reported by Native American tribes of the region.

As lamented previously, there do not seem to be any formal studies in support of this hypothesis. I have no idea where Dr. Siegel got this information; moreover, virtually every online source that features this tale either lists no source or cites back to Dr. Siegel's account. The story may well be true; we have seen stranger examples of intoxicated animals throughout this book, but I have no way to support it. I offer it to you at face value.

Another quasi-legendary account of intoxicated animals peripherally involves crop circles, circular formations that appear mysteriously in fields in various locations around the world, and which credulous sorts often attribute to aliens. In Tasmania, however, these circles have been blamed on wallabies (which look a bit like small kangaroos). About half of the world's legal opium, used to make pain-killers of various types, comes from Australian* poppy fields. In 2009, Lara Giddings, then the attorney general, and later the first woman to hold the office of premier in Tasmania, said in an interview:

> We have a problem with wallabies entering poppy fields, getting as high as a kite and going around in circles . . . Then they crash . . . We see crop circles in the poppy industry from wallabies that are high.[51]

This specific example has been observed directly and reported in news form, and the BBC quoted a representative from a poppy pro-ducer (with the straightforward name *Tasmanian Alkaloids*) as saying that while it wasn't all that common to find wallabies raiding poppy fields, they weren't the only animals that did so: "There have been many stories about sheep that have eaten some of the poppies after harvest-ing and they all walk around in circles."[52] While we don't have scientific evidence that wallabies seek out poppy plants or that, upon consuming them, they display behaviors consistent with intoxication—circular or otherwise—it at least seems plausible. Certainly, if my only choices for explaining a phenomenon like crop circles were "aliens" and "high wallabies," wallabies would get my vote every time.

## THE LEGEND OF THE ATTACK SQUIRREL

In June 2019, police officers from Limestone County in Ala-bama arrested a gentleman on drug charges.[53] Informants had told the officers that their suspect kept a squirrel in a cage, feeding it amphetamines so that it would stay aggres-sive and training it as an "attack squirrel," presumably in order

---

* Tasmania, in case you weren't aware, is an Australian state.

to sic this vicious creature on rivals and/or law enforcement. Upon arriving at the scene, police officers found said caged squirrel, but the suspect denied feeding the poor animal amphetamines. Eventually, the officers released the little guy (squirrel, not suspect) into the wild, presumably after making sure that it was not dangerous. However, this raises a question: How do you make sure that a squirrel is recovered from amphetamine-fueled attack training and is therefore not dangerous? What if the squirrel, having become addicted, manages to get its paws on some poorly guarded drugs? Do you happen to remember the 2005 movie *Hoodwinked*, featuring Twitchy, the over-caffeinated squirrel? Is there a similarly amphetamine-addled squirrel on the loose in the forests of Alabama? Have we learned nothing from Twitchy? If you ask me, this is how cryptozoological legends are born.

Now, I could not possibly leave out of this book the semi-mythical account of how animals helped us discover what became arguably the world's most widely consumed allelochemical (and the bane of Peter the Great): caffeine. It seems that people discovered the stimulating properties of coffee beans because of an observant goatherd named Kaldi, who hailed from Abyssinia (today's Ethiopia). The legend of Kaldi and his goats (Figure 6.7) is whimsically related by William H. Ukers in his 1922 book, *All About Coffee*. In part:

> *A young goatherd named Kaldi noticed one day that his goats, whose deportment up to that time had been irreproachable, were abandoning themselves to the most extravagant prancings. The venerable buck, ordinarily so dignified and solemn, bounded about like a young kid. Kaldi attributed this foolish gaiety to certain fruits of which the goats had been eating with delight. The story goes that the poor fellow had a*

*heavy heart; and in the hope of cheering himself up a little, he thought he*
*would pick and eat of the fruit. The experiment succeeded marvelously.*
*He forgot his troubles and became the happiest herder in happy Arabia.*

The fruits were, of course, coffee beans. The Muslim world is a
rich source of traditions and accounts about coffee, with probably its
most notable contribution to caffeinated culture being the name for
the beverage itesf: *qahveh*, which is pronounced exactly like its Spanish (and Portuguese and French) iteration, *café*, as well as the Italian
*caffè*, which eventually got us to the English *coffee*.

And the rest ... well, you know.

KALDI AND HIS DANCING GOATS
THE LEGENDARY DISCOVERY OF THE COFFEE DRINK
From drawings by a modern French artist

*Figure 6.7. The legendary discovery of coffee. From Ukkers (1922).*

The fourth apocryphal tale has a connection to the storied magical flying reindeer that carry that jolly old soul Santa Claus around
the world on Christmas Eve as he delivers presents to the nice and
coal to the naughty. But it begins with a very real fungus.

The mushroom *Amanita muscaria* is a beautiful species displaying white stalks supporting a white-speckled red cap (Christmassy, right?). Its two more common names are fly agaric and fly amanita. *A. muscaria* is native to temperate northern regions, but it is now found worldwide, with several subspecies that differ in coloration. The most notable characteristic of *Amanita* mushrooms is that they are all hallucinogenic, and not coincidentally, are an integral part of shamanic practices around the world.[54] Humans, shamans or otherwise, are not the only habitual consumers of *Amanita*. The relationship between reindeer and fly agaric is well known. Reindeer actively seek and eat the mushroom, and once they've done so, act as if they are drunk—sometimes *very* drunk. Interestingly, reindeer appear able to detect the smell of the chemicals in human urine; they enthusiastically consume the urine of human *Amanita* users, and can get high from that as well. In the words of Dr. Siegel, "Whenever they smell urine in the vicinity, reindeer scamper to the source and start fighting with each other for access to the clumps of yellow-stained snow."[55] (Apparently, reindeer never heard the advice about not eating yellow snow.) To be fair, it seems that reindeer are attracted to urine regardless of the source or whether the originator of said urine has consumed psychedelic mushrooms. In fact, one of the techniques that native peoples use to attract reindeer to their camps is to place strategically located buckets of urine, and apparently this works like a charm.

Reindeer behavior in response to feeding on *Amanita* or *Amanita*-laced urine is well documented in native cultures—this is not the "apocryphal" part. No, the apocryphal part is the connection between fly agaric, shamanic culture, reindeer, and Santa Claus. A controversial hypothesis proposes that *Amanita* consumption by native shamans is likely at least partially the origin of the Santa Claus story. Believe it or not, several lines of evidence seem to support this idea.[56] Christmas iconography is thought to involve a mix of Christian and pagan elements, and proponents of the psychedelic Santa hypothesis (not its actual name) point to a number of symbolic echoes between the Santa legend and shamanic practices in Siberia and other northern cultures. Reindeer were important to the indigenous peoples of Siberia, and we've already discussed the reindeer connection to *Amanita* mushrooms; there are proposals that the magical properties of Santa's deer came from humans hallucinating flying deer, or from tales of a shaman "flying high" with reindeer as companions. *Amanita* mushrooms grow under conifers—like presents under a Christmas tree—and even today Christmas ornaments in the shape of red-and-white mushrooms are common in many Scandinavian countries. Claims that this beloved character originates from shamanic traditions are far from being universally accepted,[57] but it is certainly an interesting connection to think about.

Speaking of which, it is time at last to talk about dolphins and drugs, and in doing so cover some science, some speculation, and an interesting connection between these marine mammals and the search for life in other worlds.

## THE PRIMATES OF THE SEA

Though no one would mistake a dolphin for a human, as mammals, our two species share many genetic, physiological, anatomical, and behavioral traits. And remarkably, even though the last common ancestor between the dolphin and human lineages lived about one hundred million years ago, these two different versions of "big mammals" seem to have evolved in parallel as far as social living and intelligence are concerned.[58] On account of their relatively complex brains, dolphins are widely considered to be highly intelligent, and in many ways, their neurobiology, cognitive abilities, and general behavior resemble these properties in primates.[59]

Since time immemorial, people have shared stories about the apparent nobility, intelligence, and general friendliness of dolphins.

There are numerous accounts of dolphins being friendly toward humans and toward other dolphins, even dolphins from different species. However, the evolutionary similarity between dolphins and primates goes beyond the development of intelligence and societies that exist in relative harmony. You see, the dolphin's reputation as a noble and friendly animal is somewhat . . . exaggerated. Just like us, dolphins are not always friendly or particularly noble, even toward one another. Like any species of intelligent animals, dolphin societies display a wide variability of behaviors, some of which are, sadly, a reflection of some of the least pleasant behaviors displayed by humans. Male dolphins engage in a practice known as "cooperative mate guarding" (a term that is essentially a euphemism for gang rape).[60] And dolphins show no reservations about attacking members of other non-dolphin aquatic species. Male bottlenose dolphins have been observed attacking, sexually assaulting, and killing porpoises, sometimes with considerable brutality. A hypothesis that seeks to explain this violent behavior hinges on another, particularly unsavory, behavioral practice shared by dolphins and humans: infanticide. Yes, it pains me to tell you that infanticide was extremely common in early human societies; after all, there were only so many resources to go around, and human babies are helpless and labor intensive. In bottlenose dolphins, infanticide may be related to sexual resources—dolphin calves stay with and are cared for by their mothers for a comparatively long while, and the mothers are unavailable for mating during this period; in dolphins, infanticide is usually practiced by males. To return to the porpoise attack hypothesis, adult individuals of the specific porpoise species frequently targeted by bottlenose dolphins are about the same size as a bottlenose dolphin calf. It is unclear whether dolphins confuse porpoises with dolphin younglings or practice on porpoises; either possibility is horrifying.[61] The point is that dolphins may be both less aspirational and more humanlike than is popularly assumed; they have an acute nasty bent, that's for sure.[62]

## Dolphins and LSD (or How to Talk to Aliens)

Ever since humans became humans, we have asked ourselves if there were other minds like ours out in the universe somewhere. But the search for intelligent life "out there" became both much more systematic and more "legitimate" in scientific circles in 1959, with the publication of a paper attempting to outline parameters for conducting a search for extraterrestrial intelligence (SETI), based on the best understandings of science at the time.[63] Two years later, in 1961, the Green Bank Conference on Extraterrestrial Intelligent Life was convened by the National Academy of Sciences, bringing prominent scientists from various fields together to discuss the topic. The 1961 Green Bank meeting has today attained the status of legend; among other things, it was the birthplace of the now famous Drake Equation, the first attempt at quantifying the possible number of extraterrestrial civilizations in our galaxy.*[64]

At first glance, a dolphin researcher might seem out of place at a conference about the search for extraterrestrial intelligence. The attendees of the Green Bank meetings were mostly the types of scientists you would expect: there were astronomers, like J. Peter Pearman (who came up with the idea of the meeting), Otto Struve, Su-Shu Huang, Carl Sagan (who I think requires no introduction), and Frank Drake, creator of the eponymous Drake Equation and the meeting's co-organizer. There were also engineers, as well as physicists—like Philip Morrison, one of the authors of the referenced 1959 paper, and Melvin Calvin, whose work allowed us to understand photosynthesis (in fact, he was at the Green Bank meeting when he was notified that he'd won the Nobel Prize in Chemistry), among others. However, as the story goes, once most of the attendees were agreed upon, Drake joked that "all we need now is someone who has spoken to an extraterrestrial." And in the absence of someone who spoke fluent "alienese,"

---

* In a real way, the Drake Equation does a much better job of clarifying what we *don't know* about this scientific problem; over the years, some progress has been made, but not as much as we would like.

Pearman proposed the next best thing: Dr. John Cunningham Lilly, the man who would (and thought he could) speak to dolphins.[65]

The organizers astutely reasoned that to have any hope of communicating with extraterrestrial entities, the logical approach would be to develop our abilities by trying to communicate with another intelligent species from our own planet. Dr. Lilly was invited to the Green Bank gathering because Drake had read his bestselling book, *Man and Dolphin*, published that same year, in which he detailed his ideas about the possibility for dolphin-human communication.[66] His contributions generated much enthusiasm; they even prompted the creation of an unofficial scientific society, *The Order of the Dolphin*, membership in which came with a metal pin modeled after the image of a dolphin found on an ancient Greek coin.[67]

The Green Bank meeting set the agenda for the search for extraterrestrial intelligence; an agenda that is closely followed today.* It also motivated Dr. Lilly to continue his quest to talk to dolphins. A physician and neurophysiologist with a promising scientific career, Lilly had begun his studies on dolphin behavior and communication in 1958; his research moved in the direction of human-dolphin communication when he noticed one of his dolphin subjects imitating his speech and that of a research assistant. He kept at it for almost a decade, after which his mainstream scientific career never recovered.[68] Over the years, he would claim some measure of success—a claim that was and still is widely disputed. After all, if you train a dolphin to greet you with a sound pattern similar to that of "hello," for example, what does that really mean in terms of language and communication? Lilly moved beyond trying to teach dolphins English to exploring other methods, for instance using musical tones and other nonverbal signals. His objective was to "achieve an unprecedented breakthrough beyond companionate communion to fully abstract

---

* So far, we have not yet succeeded in contacting any extraterrestrial civilization that we know of.

linguistic communication across species boundaries."[69] In other words, he wanted dolphins and humans to communicate with no more difficulty than two humans who do not speak the same language. After the Green Bank meeting, Lilly went on to publish a series of books and papers describing his dolphin research experiences. This body of work was collected in a 1975 book, *Lilly on Dolphins*, with the provocative subtitle *Humans of the Sea*.

Lilly's research would not have passed muster with current standards for animal experimentation. Along with dolphin behavior, he also studied the fundamentals of their brain biology, and in order to obtain a neurological baseline, he inserted electrodes into the brains of living, unanesthetized dolphins (anesthetized dolphins cannot breathe on their own). But even setting aside this example of outright cruelty, his approaches were . . . unorthodox, especially when it came to the methods he used (or, in some cases, allowed others to use*) to overcome the difficulties inherent in conducting behavioral research on very large aquatic animals. This is where LSD comes in.

Inspired by his personal experimentation with said chemical, he hoped to use it to help the dolphins relax and become more receptive to his communication lessons, especially those who were leery of humans because of mistreatment; one of his subjects was a rescue dolphin who had been shot three times in the tail with a spear gun. That particular dolphin is the animal referenced in the quote beginning this chapter: after being dosed with LSD, the skittish creature approached Lilly for the first time ever, regarding him in what seemed a contemplative manner. Lilly also observed that LSD increased vocalization in the dolphins, especially when another dolphin or person was present, and took this to indicate an increased desire to

---

* This aspect of Lilly's research is well documented elsewhere, but the gist is that in 1960 Lilly hired an assistant, Ms. Margaret Howe Lovatt, who developed an unusually . . . close relationship with a young male dolphin named Peter. Please see allthatsinteresting.com/margaret-howe-lovatt and theguardian.com/environment/2014/jun/08/the-dolphin-who-loved-me.

communicate. However, while he presented these findings as part of a conference on the potential uses of LSD in (human) psychotherapy,[70] they were never accepted for publication by any peer-reviewed scientific journal. Between the irregularities of method, the variability of the reactions observed, and the many possible explanations for these reactions, the usefulness of the data he collected was limited. In short, we don't really know how LSD affects dolphins, and unless we *do* eventually succeed in talking to them, we may never find out.

## Stoned Dolphins

It is worth repeating that one of the most remarkable characteristics of science is the frequency with which an apparently random observation leads to a significant scientific discovery. In 1995, while observing dolphins in the wild near the Azores Islands, a Portugal-associated archipelago in the mid-Atlantic ocean, marine biologist Dr. Lisa Steiner reported on an odd behavior in a rough-toothed dolphin, *Steno bredanensis*.[71] In her words, "One dolphin was seen pushing an inflated pufferfish along the surface." This particular observation was in no way the focus of her paper—*Steno bredanensis* was a newly discovered species, and this was a general report on the animal and what she'd seen of it. Nevertheless, this otherwise unremarkable observation established a precedent of documented interaction between dolphins and pufferfish, an interaction that relates directly to the topic of this book.

Dr. Steiner's observation of a dolphin apparently playing with a pufferfish was not the last time that such dolphin behavior was observed in the wild. Individuals of two other dolphin species, Australian humpback dolphins (*Sousa sahulensis*) and bottlenose dolphins (*Tursiops truncates*), have since been reported to "carry" or "play" with pufferfish.[72] It is well known that dolphins are curious animals that actively engage in playful behavior.[73] And it is not hard to see why a dolphin might like to play with a pufferfish, an object in its environment that puffs up and will float if you push it about—it is essentially a living (though toxic, which I'll discuss further shortly) beach ball. However, simple play does not completely explain what a group of dolphins were observed doing with a pufferfish in 2013.

Dr. Ron Pilley, a zoologist, was part of a film crew working on a documentary series about dolphins[74] when he was treated to an unexpected sight: a pod of young dolphins, passing a live (yet likely—and understandably—terrified) pufferfish around to one another. The dolphins treated the pufferfish delicately, nibbling at it and then passing it on. They evidently did not want the pufferfish for food, as in that case they would simply have eaten it. In Dr. Pilley's words: "We saw the dolphins handle the puffers with kid gloves, very gently and delicately like they were almost milking them to not upset the fish too much or kill it." He also noted that the dolphins appeared "mesmerized" by their own reflections.[75] These observations, alongside others,[76] turned on the proverbial "idea light bulb" for Dr. Pilley, who knew that the pufferfish's main claim to fame was that it produces tetrodotoxin (TTX, Figure 6.8).

## THE UNIVERSAL TOXIN[77]

One of the nastiest toxins known, tetrodotoxin is also one of a select few examples of biotoxins—toxins produced by living organisms—that are widely distributed in nature. Tetrodotoxin is found in creatures ranging from microorganisms, through otherwise apparently defenseless organisms like corals, through invertebrates like crabs, octopi, and flatworms, all the way to vertebrates like pufferfish and newts. The fact that TTX is so widely distributed in marine, freshwater, and terrestrial habitats argues in favor of its production by a variety of symbiotic bacteria or fungi—it is unlikely (though by no means impossible) that all of the aforementioned organisms share a common biochemical pathway for synthesizing tetrodotoxin. This is a small yet complex molecule that resisted all efforts to synthesize it in the laboratory until very recently (and even then it was able to be synthesized only very inefficiently). Nature has optimized the formation of TTX through a series of specialized enzymes. Current thinking about the ecological and evolutionary factors underlying tetrodotoxin's presence in organisms is that this molecule has a defensive function, and it is not hard to see why. Its function in every species in which it has been discovered so far is, well, to kill the creature's enemies. Tetrodotoxin is a paralytic toxin and has no known antidote.

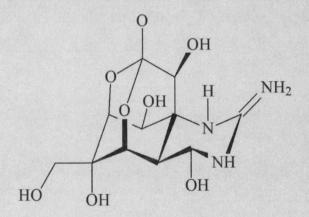

*Figure 6.8. Tetrodotoxin. Drawn by the author.*

Probably the best known of tetrodotoxin-producing organisms are various species of pufferfish. In Japan, these fish are used to prepare a traditional dish called *fugu*. For chefs preparing fugu, the aim is not to remove the tetrodotoxin entirely but to leave a small, "safe" amount; some gastronomic enthusiasts like the tingling sensation that small amounts of TTX produce in the lips and tongue (a relevant fact for our discussion). As tetrodotoxin is extremely deadly, fugu is a rather tricky dish to prepare, and can be a risky menu choice for those with plans for their future.[78]

Based on what he'd seen, Dr. Pilley hypothesized that the young dolphins were experiencing some sort of psychoactive effect as a result of being exposed to the pufferfish's tetrodotoxin. In other words, the dolphins seemed to be getting "high" on TTX.

This interesting hypothesis has been cautiously deemed worthy by various scientists in different specialties, including marine biologists, behavioral science experts, and others. However, there are also many scientists who have their doubts. Both camps make excellent points. We do not know enough of the specifics of dolphin physiology to rule out the possibility of intoxication-like effects. The "traditional" tetrodotoxin targets (sodium channels in nerve cells) have not been specifically studied in dolphins. It is not out of the question that dolphins are resistant to the toxic effects of TTX while being sensitive to possible psychoactive effects—many psychoactive drugs are, as we've seen, toxins. Dolphins, after all, have been evolving for millions of years alongside TTX-producing organisms. What's more, dolphins are much bigger than humans, meaning that they may be more resistant to the toxin. At the same time, TTX is, in my own words, a "wicked" molecule, and in fact, tetrodotoxin is a known cause of death in dolphins, despite their size[79]—it is hard to know how much toxin a given pufferfish will excrete, and if intentional, this would be an awfully dangerous form of recreation. And there is no evidence that TTX has any psychoactive effects: it does not cross the blood-brain barrier, and is bluntly a nerve toxin, which at low doses might produce tingling and even light-headedness, but beyond that progresses to paralyzing and then killing the organism that ingests it. In fact, it could be that the dolphins who appeared to Dr. Pilley to be "mes-merized" were, in fact, feeling something much less benign—perhaps they first enjoyed the curious tingling, and then accidentally enjoyed a little too much of it (Figure 6.9). The most famous scientist on the semi-skeptic side is Dr. Christie Wilcox,[80] the author of a critically acclaimed book on venoms.[81] Writing about the issue in a 2013 blog post,[82] Dr. Wilcox lays out the case against the probability of TTX having psychoactive effects, while at the same time, in true scientific spirit, keeping an open mind. As she states in her post: "I kind of *hope* I'm wrong on this one."

To be honest, I hope so too. It would make a more interesting story, and one that illustrates many of the points I've tried to make over the course of our journey.

We've seen ourselves in a number of animals throughout this book, and dolphins may be among the closest to us of those we've discussed. At the very least, they are mentally sophisticated enough to be spoken of in terms of psychology, which means that their motivations are, in a way, harder to discern than those of more "instinctual" creatures. Who's to say whether they are getting "high," intentionally or otherwise, or are simply beguiled by the tingling sensation that nibbling pufferfish can induce—just like that many of our fellow humans enjoy while eating fugu, that most dangerous of Japanese delicacies?[83]

## CODA

My purpose in writing this book was to share with you some of the fascinating behaviors that humans and a variety of organisms on this earth display in seeking and consuming psychoactive and medicinal substances. It is an undeniable fact that these behaviors are a direct consequence of our shared evolution, and for that reason alone, they are both instructive and remarkable.

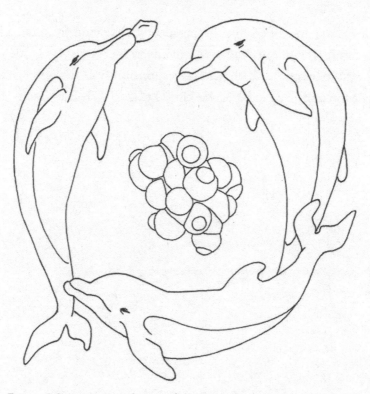

*Figure 6.9. Artistic rendering of three dolphins passing a pufferfish to one another, with a three-dimensional representation of tetrodotoxin in the middle. Drawing by Ms. Alisa Singh.*

Throughout this book we saw that seeking and consuming psychoactive substances is a drive that is an integral part of our biological makeup, yet the fact that drug use is natural (and make no mistake, it is) does not mean it is universally beneficial or benign, or even inevitable.

Borrowing an idea originally expressed by Dr. Richard Dawkins at the end of his first (and still his best) book,[84] genetics need not become destiny. I neither advocate nor judge drug use, much less abuse; I will only say that humans are endowed with what seems to be a unique capacity to understand and use the chemical gifts of nature, as well as the capacity to study and even resist the control of a series of desires and instincts that served us well for evolutionary purposes in times long gone. Perhaps the perspectives we gain from becoming more familiar with our fellow organisms in this book can guide us in our quest to better fulfill our sacred responsibility as the only incarnation of the universe (as far as we know) that can observe itself.

# ACKNOWLEDGMENTS

Once again, dear reader, thank you for choosing to read my work; this is the bottom line for every author since the beginning of time. You will never know how much it means to me.

Now I wish to thank some of the wonderful souls who helped me make this book adventure a reality.

First, I want to express my deepest appreciation to Mr. Glenn Yeffeth, publisher of BenBella Books. Thanks, Glenn, for giving me the opportunity to develop my popular science writing.

I also want to thank the whole BenBella team, but especially those members of the team with whom I worked directly: Scott Calamar, Michael Fedison, Jennifer Canzoneri, Sarah Avinger, Alicia Kania, Adrienne Lang, Monica Lowry, Jay Kilburn, and Leah Wilson. It is always a pleasure working with such a talented and professional team; thank you one and all. There are two additional BenBella team members I want to thank for their invaluable help, writing advice, guidance, and patience. They are my two Jedi Masters, and they were my editors for this book: Laurel Leigh and Alexa Stevenson. Alexa and Laurel, just as with *Strange Survivors*, what can I say? Words are still woefully inadequate to convey my gratitude. This book is much better because of your advice and gentle nudging; I am a better writer and found my "voice" because of you. Simply stated: Thank you, and I hope I make you proud!

I want to thank my wife, Liza, for her help, suggestions, and above all, her love. She gave me her keen eye and her peerless common sense to make this book even better. She knows how to slow me down and keep me grounded in those times when I tend to go "full science ahead." Also, I would be remiss if I didn't acknowledge our (first) twenty-nine years together. I love you, my dream girl!

My children, Vanessa, Reynaldo, and Andy: Because of you, I try to be a better man and a better dad every day of my life. I live for your love, your hugs, your smiles, and for your encouragement and understanding. You make me proud each and every day, and it is an honor and a privilege to be your dad. I love you with all my heart.

I cannot pass up this opportunity to express my gratitude to the small yet powerful army of talented students I have had the privilege of working with over the last sixteen years (as of this writing, almost eighty or so students and counting), who have helped my research program grow over time. Virtually all of these bright, inquisitive, and hardworking students have gone on to pursue productive careers of their own. I am very—and I mean *very*—proud of the fact that there are people out there in the world doing scientific research, treating patients, teaching, and engaging in many other activities that positively contribute to society, who got their start by acquiring research experience in my laboratory. Also, I wish to thank my senior seminar group of spring 2018, which had a lot of fun talking with me about animals and drugs.

A good library is the happy place of a good scholar. I have said this before, and I am saying it again. The library of my academic home, West Chester University (WCU), was invaluable when looking for information in many forms: books, original papers, electronic resources, and other materials. I want to point out that this book was finished in the midst of the COVID-19 pandemic, and the way in which the remote resources of the library were managed and kept working flawlessly was second to none. This is a testament to the great staff of the WCU library and especially to my friend and colleague Dr. Walter Cressler, who is the library's "go to" guy as far as science is concerned. Thanks, Walt!

I wish to thank Dr. Barbara J. Cordell, Dr. Jeffrey D. Sommer, Mr. Gary Ray Rogers, and Dr. Charles I. Abramson for kindly sending reference material that I would not have been able to get otherwise. Aside from the actual writing, I had the most fun finding illustrations for the book, and I wish to thank the following friends and colleagues for helping me out: Dr. Ulrike Heberlein (Figure 1.2), Dr. Michael Smith (Figure 5.1), Ms. Chelsea Linaeve (Figure 5.3), Dr. Robert (Bób) Raffa (Figure 5.4), Dr. Danna Staaf (Figure 5.5), Ms. Alisa Singh (Figure 6.9), and very especially Mrs. Elizabeth M. Rivera (a.k.a. Mrs. Baldscientist, my better—and prettier—half), and Ms. Giselle Vanessa Pagán (a.k.a. my baby girl), for Figures 5.2 and 6.4, respectively.

I want to express my special thanks to Matthew D. LaPlante, Peter Cawdron, Sofía Villalpando, Marie McNeely, Justin Gregg, and Micah Hanks for your endorsements. I am humbled and honored by your kindness.

Finally, any errors in the book are mine and mine alone! (And, along similar lines, if I forgot anyone in these acknowledgments, my apologies.)

# NOTES

## Author's Note

1   For an excellent general book about the effects of abused drugs in society and an excellent primer on what these types of drugs are in the context of brain science, please see Grisel (2020) *Never Enough: The Neuroscience and Experience of Addiction.*

2   Please see Siegel (2005).

3   Please see Samorini (2002).

## Introduction

1   Please see nytimes.com/1981/04/17/movies/caveman-with-ringo-starr.html.

2   Actually, this research was published by the group of the world's leading authority on the paleontological applications of amber, Dr. George Poinar, currently affiliated with Oregon State University. Dr. Poinar has another claim to fame: he was the person who suggested that dinosaur DNA could be extracted from bloodsucking insects preserved in amber. This idea of course led to the plot of the famous *Jurassic Park* books by the late Michael Crichton, and the subsequent movies.

3   Please see Poinar and collaborators (2015). For a nice summary of this research, please see portlandmercury.com/BlogtownPDX/archives/2015/06/10/dinosaurs-disease-and-drugsand-the-scientist-behind-jurassic-park.

4   Please see Prasad and collaborators (2005).

5   For a reader-friendly, comprehensive review of psychoactive animals, please see Orsolini and collaborators' 2018 review "Psychedelic Fauna for Psychonaut Hunters" in *Frontiers in Psychiatry.*

## Chapter 1

1   Throughout this book I will generally give an animal's common name followed by its scientific name. For a short illustration of why it is useful to use scientific names, please see Pagán (2018), pages 10–11. For some general articles about koalas, please see media.australianmuseum.net.au/media/dd/Uploads/Documents/27761/Koala+fact+sheet+May+2014.ea2c198.pdf and www.mentalfloss.com/article/59114/10-things-you-didnt-know-about-koalas.

2   Charles Darwin (of "evolution" fame) was one of the best-known naturalists of the eighteenth century—scratch that—of *any* century. I am assuming that you, my faithful reader, are familiar with the biological evolution concept and its most famous exponent, so I will not elaborate much about that in this book. However, if you need a refresher, or if you want to know more, here are some suggestions: Pallen (2009), *The Rough Guide to Evolution*; Dawkins (1995), *The River Out of Eden: A Darwinian View of Life*; and my own: Pagán (2018), *Strange Survivors: How Organisms Attack and Defend in the Game of Life.* I expressed some of my thoughts here: baldscientist.com/2014/08/22/charles-darwin-blogger/.

3   I found many such letters (including Nicols's) in a thorough collection of Darwin's correspondence (www.darwinproject.ac.uk/). Another excellent source of Darwin's writings is darwin-online.org.uk/. I actually got this quote from darwinproject.ac.uk/.

4   *From So Simple a Beginning: Darwin's Four Great Books* (Voyage of the Beagle, The Origin of Species, The Descent of Man, The Expression of Emotions in Man and Animals), Edward O. Wilson, editor, p. 785.

5   Alfred Edmund Brehm, a zoologist and explorer, who was the first director of the Zoological Garden of Hamburg.

6   From *So Simple a Beginning*, p. 785.

7   Please see Siegel (2005), Chapter 5.

8    Please see Zuckerman and collaborators (1975).

9    Give or take; please see Keller (1979) and Vallee (1998).

10   Please see Braidwood and collaborators (1953).

11   For a really nice summary of these ideas, please see Dominy (2015).

12   Please see Rasmussen (2014) and Vallee (1998).

13   The tale of humans and alcohol is more complex and nuanced than we are able to fully explore in this book; therefore, I refer you to a few excellent sources should you want to know more about this interesting story. Please see Dominy (2015), Goode (2014), McGovern (2009), Rasmussen (2014), and Vallee (1998).

14   Please see Dashko and collaborators (2014).

15   For a brief exploration of the evils of oxygen, please see pages 84–87 of my *Strange Survivors* (Pagán, 2018).

16   Please see Auesukaree (2017), Albergaria and Arneborg (2016), Goode (2014), and Snoek and collaborators (2016).

17   Please see Hernández-Tobías and collaborators (2011).

18   Please see Yoshida and collaborators (1998).

19   Please see Edenberg (2007), Edenberg and McClintick (2018), and Hjelmqvist and collaborators (2003).

20   Please see Cordell and collaborators (2019), Cordell and Kanodia (2015), Kaji and collaborators (1976), and Painter and Sticco (2019).

21   From Ladkin and Davis (1948), and from Sato (1952), as referenced by Guo and collaborators (2019).

22   Please see Cordell and McCarthy (2013) and Malik and collaborators (2019).

23   Please see Joneja and collaborators (1997) and Cordell and collaborators (2019).

24   Please see Logan and Jones (2000) and cnn.com/2015/12/31/health/auto-brewery-syndrome-dui-womans-body-brews-own-alcohol/. For a layperson-friendly yet comprehensive treatise of ABS, please see Dr. Barbara Cordell's 2019 book, *My Gut Makes Alcohol!: The Science and Stories of Auto-Brewery Syndrome*, and the article "*Candida albicans*, the Yeast Syndrome, and the Auto-Brewery Syndrome" at voiceforthedefenseonline.com/newsletters/2017/May2017.pdf.

25   Please see theatlantic.com/science/archive/2019/09/drunk-without-alcohol-autobrewery-syndrome/598414/. For a more clinical description of this syndrome, please see ncbi.nlm.nih.gov/books/NBK513346. For more clinical reports, refer to Akhavan and collaborators (2019) and Guo and collaborators (2019). Also, see Kruckenberg and collaborators (2020), and for a popular version of this story, usatoday.com/story/news/nation/2020/02/25/auto-brewery-syndrome-woman-didnt-drink-but-urine-full-alcohol/4866137002/.

26   Please see Hafez and collaborators (2017).

27   Please see Bivin and Heinen (1985).

28   Please see van Waarde (1990).

29   Please see Fagernes (2017). For a reader-friendly summary of this work, please see smithsonianmag.com/smart-news/now-we-know-how-goldfish-produce-alcohol-180964502/.

30   From his October 21, 2019 radio show/podcast episode.

31   Please see Janecka and collaborators (2007), and Li and Ni (2016).

32.  Please see Gochman and collaborators (2016).

33   Please see Wiens and collaborators (2008).

34   This is one of the ways in which alcohol can be metabolized. It is called *glucuronidation*, and happens when a glucose derivative (glucuronic acid) combines with a chemical (in this case alcohol) and makes said chemical easier to eliminate in urine. Interestingly, tree shrews are indeed catching the attention of researchers interested in drug addiction. There are studies aimed to the establishment of this species as an animal model of morphine effects. Please see Shen and collaborators (2014).

35   If you are curious about this interesting episode in the history of biology, please see Pagán (2014) *The First Brain*. Also, for a very readable historical overview of this genus, please see O'Grady and DeSalle (2018). If you want to know more about the role that the fruit fly played in the fascinating story of modern genetics and developmental biology (with some behavioral studies added for taste), please see Brookes (2002) *Fly: The Unsung Hero of Twentieth Century Science* and Weiner (2000) *Time, Love, Memory: A Great*

*Biologist and His Quest for the Origins of Behavior.* For a more recent update on *Drosophila* science, please see Mohr (2018) *First in Fly: Drosophila Research and Biological Discovery.*

36 Please see Peterson and collaborators (2008).

37 For two excellent reviews of drug responses and addiction in *Drosophila*, please see Lowenstein and Velazquez-Ulloa (2018) and Ryvkin and collaborators (2018). Also, there is a reader-friendly review very appropriately titled "Insights from intoxicated *Drosophila*"; please see Petruccelli and Kaun (2019).

38 *Pomace* is the remainder of fruits (pulp) after being pressed to extract their juice. Please see Ojelade and Rothenfluh (2009).

39 Please see Ursprung and Carlin (1968).

40 For more details about beer and wine traps, please see Ni and collaborators (2008), Pettersson and Franzén (2008), Laaksonen and collaborators (2006), and Vassiliou (2011).

41 This was expertly reviewed in Sekhon and collaborators (2016).

42 Please see Fry (2014).

43 For representative examples of a variety of inebriometer designs and how scientists used them, please see Berger and collaborators (2004), Dawson and collaborators (2013), Ojelade and Rothenfluh (2009), Singh and Heberlein (2000), and Wolf and Heberlein (2003).

44 Please see Lindsay (1879), which he dedicated to his late father in a very affectionate way: "Who, of all my many correspondents, was surpassed by none, either in his generous, genial, and genuine sympathy with the domestic animals, or in the sound, liberal, and philosophical views which he held and expressed regarding the nature and extent of animal reason." File downloaded from archive.org. I heard of this specific work of Dr. Lindsay's from chapter five of Dr. Siegel's 2005 book, *Intoxication.*

45 Siegel, *Intoxication, The Universal Drive for Mind-Altering Substances.*

46 Please see Firn (2005), pages 31–35 and 41–46.

47 As narrated by Vallee (2008).

48 Not everybody agrees with the characterization of addictions as a disease. Please see Lewis (2015), and Siegel (2005b) in McSweeney, Murphy, and Kowal (2005).

49 *Drugs, Brains, and Behavior: The Science of Addiction,* by the National Institute of Drug Abuse (NIDA, 2018), although I learned of this formal definition for the first time in Lewis (2016).

50 This is an area that is under furious investigation, with exciting discoveries published every single week. Some very accessible books that tell in more detail how animal models are used in addiction research include Samorini (2002) and Siegel (2005). For more technical reviews, please see Ahmed (2012), Cates and collaborators (2019), García-Pardo and collaborators (2017), Müller (2018), Pagán (2005, 2014, 2017, 2019), Søvik and Barron (2013), Spanagel (2017), Wolf and Heberlein (2003). A few excellent sources should you want to learn more about addiction in general include Erickson (2018) *The Science of Addiction: From Neurobiology to Treatment,* Kuhar (2015) *The Addicted Brain: Why We Abuse Drugs, Alcohol, and Nicotine,* and Lewis (2016) *The Biology of Desire: Why Addiction Is Not a Disease.*

51 For an excellent and reader-friendly overview of dopamine's role in this regard, please see Lieberman and Long (2018) *The Molecule of More: How a Single Chemical in Your Brain Drives Love, Sex, and Creativity—and Will Determine the Fate of the Human Race.*

## Chapter 2

1 Dr. Lewin's story is rather poignant. Despite being an impeccably credentialed scientist and a gifted teacher and researcher, he never got a permanent academic position at a university. Dr. Bo Holmstedt, of the department of pharmacology at the Karolinska Institute, speculated in his preface to *Phantastica*'s reprint edition that this was likely due to anti-Semitism.

2 However, there are some interesting pieces of evidence suggesting that nicotine can act as a neuroprotective agent in animals, as well as a possible pharmacotherapy against Parkinson's disease (also done only in animals so far). For representative publications, please see Ferrea and Winterer (2009), Thiriez and collaborators (2011), and Quik and collaborators (2008).

3 In general, "dose-response" refers to the administration of drugs to humans or animals, while "concentration-response" refers to the testing of drugs in tissue, cells, or biochemical preparations. Please see Tsatsakis and collaborators (2018).

4 For an excellent exploration of the dose-response concept within the context of toxicology, please see Calabrese (2016).

5 Please see my book *Strange Survivors: How Organisms Attack and Defend in the Game of Life,* Pagán (2018), for a more detailed exposition of evolutionary processes.

6 The best available historical evidence dates this companionship to about six thousand years or so, but again, recorded history is usually "younger" than "actual" history. Please see Brook and collaborators (2017). There is archaeological evidence that indicates that Stone Age peoples did cultivate poppy plants (please see Siegel, 2005, chapter five).

7 The intertwined history of opium and humanity is a fascinating topic. Please see Grover (1965), Schiff (2002), and Siegel (2005), chapter five, for a comprehensive historical account of opium use and abuse.

8 From Schiff (2002).

9 The economic consequences of cocaine abuse are widely documented; on the other hand, the social effects of such activities are equally documented, but very difficult to quantify. For a couple of thorough reviews of the societal and economic effects of cocaine abuse, please see Frazer and collaborators (2018) and Schneider and collaborators (2018).

10 I would ordinarily ask you to go "see" the reference, but it is unlikely that this manuscript would be available at your local library, so here's the complete reference: Cobo, B. (1653) Historia del Nuevo Mundo. Manuscrito en Lima, Perú, 1653, libro 5°, capítulo XXIX, as cited in Calatayud and González (2003). I narrated a version of the story of Father Cobo in Pagán (2014).

11 Please see Pagán (2014), chapter five.

12 Cocaine's best-understood pharmacological targets in terms of its local anesthetic effects and addictive properties, respectively, are a particular type of voltage-gated sodium channel and certain neurotransmitter transporters (please see Pagán, 2005, and Pagán, 2014, chapter five). Before we knew of the molecular basis of these effects, though, all we knew about khoka leaves at first was that they seemed to be useful for alleviating pain, fatigue, and so on. Later on, once we identified the chemical responsible for these effects (cocaine), the specific reasons for these effects became much better understood.

13 Please see Markel (2011) "Über Coca: Sigmund Freud, Carl Koller, and Cocaine."

14 The idea of natural selection occurred independently to Alfred Russell Wallace, an interesting story that is very well explored elsewhere. Please see Browne (2013) for a concise yet pretty thorough account of this episode in the history of evolutionary thought. Charles Darwin's grandfather, Erasmus Darwin, in addition to being a physician, was a distinguished naturalist in his own right, who came up with what can be considered early evolutionary ideas.

15 And this does not surprise me in the least; please see Antolin (2011) and Hayman (2009). Darwin also tried to train for a career as a clergyman, but he dropped out of that as well.

16 Please see Markel (2011).

17 I must say that I got very excited when I learned about the "Coca Koller" nickname, and I am sure I was not the first to connect it with what arguably is one of the most famous products that—used to—contain cocaine: Coca-Cola. Alas, Freud was not part of the "origin story" of the Coca-Cola name. The name came from Frank Robinson, a marketing expert who was an associate of pharmacist John Pemberton, the actual inventor of what to was to be an earlier version of the famous beverage. Robinson coined the Coca-Cola name from (of course) the coca leaf and the kola nut extract used in the recipe. (I thank Mr. Mark Pendergrast, author of the book *For God, Country, and Coca-Cola: The Definitive History of the Great American Soft Drink and the Company That Makes It,* for graciously providing this information via email.) At the end of the day, since Freud was more famous than Koller, the "popularization" of cocaine was and still is generally—and unjustly—credited to Freud.

18 Incidentally, Coca-Cola was not the only drink that had coca leaves as the main ingredient. In the 1800s, Angelo François Mariani, a pharmacist, began an association with a physician, Charles Fauvel, one of the first doctors to use cocaine as an anesthetic in nose and throat surgery. Fauvel's sponsorship of Mariani's apothecary was due in no small part to Mariani's interest in developing a coca-based drink. The product was eventually called Vin Mariani, a rather successful coca-based wine. The success that Vin Mariani enjoyed was not so much due to the quality of the product as much as it was due to Angelo's excellent marketing campaign, particularly by presenting the product as a medicinal tonic, which caught the eye of

the physicians of his day. This is a very interesting story masterfully told elsewhere (please see Helfand, 1980, and Stolberg, 2011).

19   For a more extensive narration of this story, see Markel (2011).

20   Please see Fishman (2011).

21   In Mantegazza (1858) *Sulle virtù igieniche e medicinali della coca e sugli alimenti nervosi in generale*, and Mantegazza (1859) *Sull'introduzione in Europa della coca, nuovo alimento nervosa*, as reported by Samorini (1995).

22   Please see Siegel (2005), page 170, or Boucher and Moeser (1991), Knuepfer (2003), and Van Dyke and Byck (1982).

23   For more information on the idea of using this insect against coca plants, please see "*Eloria Noyesi*: Colombia's Potential Solution to Eradicating Illicit Coca": coha.org/eloria-noyesi-colombias-potential-solution-to-eradicating-illicit-coca/.

24   Historically, cocaine is one of the main secondary metabolites isolated from some fifteen of about two hundred known but otherwise unremarkable plant species belonging to the *Erythroxylon* genus that live in tropical climates. This plant genus has a very interesting evolutionary history (please see Islam, 2011). Of these fifteen *Erythroxylon* species, there are two species in particular that produce the highest concentration of cocaine (about 1.8 percent by leaf dry weight): *E. coca* and *E. novogranatense.* (Please see Calatayud and González, 2003, and Boucher and Moeser, 1991.)

25   Please see Blum and collaborators (1981).

26   Please see Chen and collaborators (2006).

27   For two nice reviews of the natural history and medicinal chemistry of cannabinoids, please see McPartland (2018), McPartland and collaborators (2006), and Vemuri and Makriyannis (2015).

28   Please see Breivogel and collaborators (2018) and McPartland and Glass (2003).

29   Please see Buttarelli and collaborators (2002), McPartland and collaborators (2006), and Rawls and collaborators (2008).

30   Please see McPartland (2004), McPartland and collaborators (2001), and McPartland and Glass (2003).

31   Please see Jimenez-Del-Rio and collaborators (2008), McPartland (2004), and McPartland and collaborators (2006a, b).

32   Please see da Fonseca and collaborators (2016), Emerich and collaborators (2016), and Freitas and collaborators (2016).

33   For example, please see Flicker and collaborators (2019).

34   See Siegel (2005), p. 158

35   From Nichols (2016).

36   From Jaffe (1990).

37   Please see Osmond (1957) "A review of the clinical effects of psychotomimetic agents."

38   A delightful account of the discovery of LSD (*LSD: My Problem Child*) was penned by Dr. Hofmann (please see Hofmann, 2009).

39   Hofmann (2009), p. 15.

40   For excellent reviews on the evolution of *Claviceps* fungi, please see Florea and collaborators (2017) and Píchová and collaborators (2018).

41   For a thorough review of ergotism and its symptoms, please see Belser-Ehrlich and collaborators (2013).

42   Please see McKenna (1992) *Food of the Gods: The Search for the Original Tree of Knowledge. A Radical History of Plants, Drugs, and Human Evolution*, and Haarmann and collaborators (2009).

43   Please see Haarmann and collaborators (2009) and Lorenz and collaborators (2009).

44   Please see McKenna (1993) and Wasson and collaborators (2008) "The Road to Eleusis: Unveiling the Secret of the Mysteries."

45   Please see Kalan and collaborators (2019), and Kühl and collaborators (2016).

46   Siegel (1977).

47   By the way, the stoned ape hypothesis is colloquially termed the stoned ape *theory*, but this latter is a misnomer, because this is not a scientific theory; it is actually a hypothesis. For a concise explanation of the scientific meanings and differences between theories and hypotheses, please see my book *The First Brain* (Pagán, 2014), pages 12–14.

48  Please see Nichols (2014, 2016) and Winkelman (2017).

49  Please see McKenna (1992) *Food of the Gods: The Search for the Original Tree of Knowledge. A Radical History of Plants, Drugs, and Human Evolution.*

50  Please see Rightmire (2004), Shultz and collaborators (2012), and Sloat (2017 at inverse.com/archive/july/2017/science).

51  Again, please see Sloat (2017 at inverse.com/archive/july/2017/science). Also, please see newscientist.com/article/mg21128311-800-a-brief-history-of-the-brain/ and theatlantic.com/science/archive/2018/03/a-deeper-origin-of-complex-human-cultures/555674/.

52  Please see Dorus and collaborators (2004) and Fu and collaborators (2011).

53  McKenna (1992) *Food of the Gods: The Search for the Original Tree of Knowledge. A Radical History of Plants, Drugs, and Human Evolution.*

54  de las Casas B (1875) Historia de las Indias, Madrid, Impr. de M. Ginesta.

55  For a general review of various antiparasite strategies used by birds, please see Bush and Clayton (2018).

56  Please see Suárez-Rodríguez and collaborators (2012).

57  Please see Suárez-Rodríguez and García (2017).

58  Please see Suárez-Rodríguez and García (2014).

## Chapter Three

1  Please see Solecki (1975). Also, for a delightful book with more detailed information about the Neanderthals, please see Papagianni and Morse (2015) *The Neanderthals Rediscovered.*

2  It seems that the DNA of certain African populations contain up to 0.3 percent of Neanderthal DNA. This research is still unpublished, but an informal summary of these findings can be found here: https://science.sciencemag.org/content/367/6477/497. To learn more about the genetic legacy that we share with Neanderthals, please see Wall and collaborators (2013).

3  Please see Rogers and collaborators (2020).

4  For an excellent summary of these points, please see Shipley and Kindscher (2016). Also, please see arstechnica.com/science/2018/02/neanderthals-were-artists-and-thought-symbolically-new-studies-argue/.

5  Please see Hoffmann and collaborators (2018a, b, c), García-Diez and collaborators (2015), and Pike and collaborators (2012).

6  Please see Lietava (1992), Leroi-Gourhan (1975), and Merlin (2003).

7  Please see sciencemag.org/news/2019/01/new-remains-discovered-site-famous-neanderthal-flower-burial.

8  Reported in detail in Hardy and collaborators (2012, 2018) and in Weyrich and collaborators (2017). Also, please see Maderspacher (2008).

9  Please see Grienke and collaborators (2014), Vunduk and collaborators (2015), and Yun and collaborators (2012).

10  I recently came across a very reader-friendly and concise introduction to the concept of consciousness; please see Harris (2019). Also, please see Blackmore (2017) and Pagán (2019).

11  Please see Musgrave and collaborators (2016, 2019), Pascual-Garrido (2019), and Whiten (2017).

12  Please see Hölldobler and Wilson (2008, 2010), Pagán (2018), chapter six, and Pagán (2019).

13  If you want to know more about this interesting organism, please refer to my *Strange Survivors*, page 123.

14  But there are very well-known (and very interesting) cases of humans who use echolocation due to acquired or innate blindness; please see Flanagin and collaborators (2017), Thaler and collaborators (2019), and Thaler and Goodale (2016).

15  Nagel (1974). Incidentally, in 2012, Dr. Nagel wrote an interesting yet controversial book that criticized current evolutionary paradigms in light of their apparent inefficiency at explaining mind and consciousness. Please see *Mind & Cosmos: Why the Materialist Neo-Darwinian Conception of Nature Is Almost Certainly False.* There is no such thing as a "retired" scholar.

16  For a delightful exploration of animal emotions, please see Laurel Braitman's 2015 book, *Animal Madness: Inside Their Minds.*

17  Some examples of clear, objective review articles in this topic include Anderson and Adolphs (2014), Adolphs and Anderson (2013), Baracchi and collaborators (2017), Bateson and collaborators (2011), Mather (2011), and Mendl and collaborators (2011).

18  For two in-depth explorations of the minds of animals, please see Griffin (1981, 2001). For two additional (and more reader-friendly) sources, please see de Waal (2017) and Safina (2016). Darwin said these prescient words in his 1871 book, *The Descent of Man*. Finally, for a thorough yet reader-friendly exploration of animal feelings and consciousness, please see Tye (2016) *Tense Bees and Shell-Shocked Crabs: Are Animals Conscious?*.

19  But we *do* know what happens when we give cocaine to certain aquatic snails; please see Carter and collaborators (2006).

20  Please see de Pasquale (1984) for a longer narrative on the history of pharmacognosy.

21  For further reading on the influence of other scientific fields on pharmacognosy, please see Larsson and collaborators (2008). A detailed account of the history of pharmacognosy is beyond the scope of this book, and therefore I refer you to some excellent and more detailed narratives of this history. Please see de Pasquale (1984) and Heinrich and Anagnostou (2017).

22  Obviously, using the pharmacognosy practice of humans as an inspiration. Please see Rodríguez and Wrangham (1993).

23  As narrated by Campbell and Rodríguez (1996).

24  From Huffman (1997) and paraphrased from Shurkin (2014).

25  Please see Samorini (1995).

26  As narrated in Huffman (2001).

27  Please see Janzen (1978).

28  As reported by Dr. Michael Boppré (1984).

29  From Boppré (1984). For a very good review of pharmacophagy, please see Abbott (2014).

30  Please see Pagán (2018), pages 122–123.

31  As defined in 2011 by Young and collaborators in an article whimsically titled "Why on Earth?"

32  Please see Fox (1971). In many cases, certain practices unfamiliar to Western science are simply labeled as "pathological." It is only through further study that in many cases the ubiquity of a certain behavior is established.

33  Reviewed in Henry and Cring (2013).

34  Please see Raman and Kandula (2008).

35  Reviewed in Young (2011).

36  This "list" also includes bison, rhinos, pigs (domestic and wild), and so on. Please see two excellent and detailed reviews: Engel (2002) and Huffman (2003).

37  Please see Abbott (2014); for a more technical exploration of phenotypic plasticity in the context of animal self-medication, please see Choisy and de Roode (2014). Also, we'll see how fruit flies use alcohol for medicinal purposes in chapter five.

38  A discussion of the monarch's mass migrations and their sophisticated navigation strategies and mechanisms is beyond the scope of this book, but you may want to check out a few excellent and readable references. Please see Reppert and de Roode (2018) and Reppert and collaborators (2010, 2016).

39  Named by the father of taxonomy himself, Linnaeus, after the Greek god of healing, *Asklepios*, on account of the plants' medicinal properties. For a comprehensive review of this plant and its interaction with monarchs, please see Rasmann and Agrawal (2011).

40  Please see Petschenka and collaborators (2015, 2016, 2017, 2018). I talked more extensively about this protein in my book *The First Brain* (Pagán, 2014), pages 51–53.

41  For an early and very readable article on this topic, please see Brower and Glazier (1975).

42  For an excellent review on this topic, please see Hoang and collaborators (2017). Also, please see Tan and collaborators (2018).

43  Trans-generational immunity is a phenomenon well described in vertebrate organisms, but only discovered in insects in 2006. Please see Pigeault and collaborators (2016), Tetreau and collaborators (2019), and Roth and collaborators (2018).

44  Please see Sternberg and collaborators (2015).

45    Please see Jones and collaborators (2019).

46    Please see Agrawal and Hastings (2019) and Tao and collaborators (2016).

47    For a comprehensive story of the monarch/milkweed link, I refer you to an excellent general book on the topic: Please see Agrawal (2017) *Monarchs and Milkweed: A Migrating Butterfly, a Poisonous Plant, and Their Remarkable Story of Coevolution.*

48    Please see Soloway (1976) and Steppuhn and collaborators (2004). Incidentally, tobacco plants are not the only plants that can make nicotine. Most members of the *Solanacea* family (tomatoes, potatoes, broccoli, eggplant, etc.), among other plants, make small yet detectable amounts of nicotine. Please see Domino and collaborators (1993). Also, please see "Nicotine Keeps Leaf-Loving Herbivores at Bay," DOI: 10.1371/journal.pbio.0020250.

49    As described by McArt and collaborators (2014).

50    Please see Baracchi and collaborators (2015) and du Rand and collaborators (2017).

51    Please see Anthony and collaborators (2015), Richardson and collaborators (2015), and Thorburn and collaborators (2015).

52    A term coined by the science writer and entrepreneur Janine Benyus; please see Benyus (2002) *Biomimicry: Innovation Inspired by Nature.*

53    I don't want to digress too much and stray from our story (and I know I will!); therefore, please see Alvarez and collaborators (2021), Ibrahim and collaborators (2021), and Ng and collaborators (2020). Also, for a general introduction to biomimetic applications, please see the delightful book *Biomimicry: Innovation Inspired by Nature* (Benyus, 2002).

54    Please see Groot and collaborators (2016).

55    The actual figure is 479 million years. Please see Misof and collaborators (2014).

56    For a more extensive review of these studies, please see Rains and collaborators (2008) and Schott and collaborators (2015).

57    In fact, in a previous work, I described superorganisms as "brains within brains" (Pagán, 2019).

58    The nature and character of superorganisms is beyond the scope of this book, but the fact that community-wide self-medicating and cooperative behaviors do exist promises to give us fascinating insights on the awe-inspiring complexity of biological life on Earth. For two excellent general references on eusociality, please see Hölldobler and Wilson (2008) *The Superorganism: The Beauty, Elegance, and Strangeness of Insect Societies*, and Hölldobler and Wilson (2010) *The Leafcutter Ants: Civilization by Instinct.* For a couple of more specialized works, please see Wilson and Hölldobler (2005) and Queller and Strassmann (2003), and if you have time, take a look at my 2018 book, *Strange Survivors.*

59    Please see Simone-Finstrom and Spivak (2012) and Simone-Finstrom and collaborators (2009, 2017).

60    Please see Evans and Spivak (2010) and Spivak and collaborators (2019).

## Chapter Four

1    Please see Hedges and collaborators (2004, 2015, 2018), Kumar and collaborators (2017), and Wang and collaborators (1999). Surprisingly, the most recent data available seems to indicate that animals and fungi diverged *after* they split from plants, roughly 1.2 billion years ago. I used the TimeTree project website (timetree.org) to calculate these timescales. This is an excellent and user-friendly resource in case you might be interested on the evolutionary relationships between organisms and the timescales involved.

2    The formal notion of plants as supporting characters began with (who else?) the ancient Greeks. More or less at the turn of the fifteenth century (but really gaining traction by the seventeenth), some naturalists began to think of plant life as more than background noise; these thinkers—quite fittingly—found plants fascinating, and as a consequence they began to systematically study plants with gusto despite a relative lack of sophisticated technology. This fact is quite remarkable, since even with the technological limitations of their time, many ingenious naturalists came up with clever ways to tease from nature some of the secrets of how plants work. One of these clever scholars was Charles Darwin, who first by himself and later in collaboration with his son Francis (who was a distinguished scientist in his own right and an actual plant biologist) produced a large and significant body of work on plants, particularly on aspects of their behavior. As a testament to their observational and scientific capabilities, a significant proportion of the observations and discoveries the Darwins made about plants, as well as their interpretation of those

findings, are still valid today, almost a hundred and fifty years later. But with the notable exception of the Darwins, plant life was relatively invisible to most, whether they were laypeople or professional educators in colleges and universities. (For a series of very readable reviews of the Darwins' contribution to plant science, please see Baluska and collaborators [2009]; Kutschera and Briggs [2009]; Kutschera and Niklas [2009]; and Hopper and Lambers [2009]. For a short survey on other naturalists who studied plants, please see Gagliano [2013].)

This state of affairs persisted as late as the beginning of the twentieth century and was lamented by a variety of scholars. (Please see Nichols [1919].) This apparent invisibility likewise puzzled a variety of scholars, and this sparked efforts to determine *why* plants attracted less interest from laypeople and scientists alike. In 1998, two botanists and educators, James Wandersee and Elizabeth Schussler, proposed the term *plant blindness* to promote awareness of the prevalent lack of interest in plant life, particularly in developed countries. (Please see Wandersee and Schussler [1999].) They defined plant blindness as the "inability to notice plants in their environment and a failure to recognize and appreciate the utility of plants to life on Earth coupled with a belief that plants are somehow inferior to animals." (From Balas and Momsen [2014]; also, please see Allen [2003].) Wandersee and Schussler's phrasing strongly implied that people's attitudes about plants went beyond a mere preference for animals over plants, an attitude sometimes called *zoocentrism*. Zoocentrism implies that plants elicit less curiosity and attention from us than animals simply because we see ourselves reflected in animals (it kind of makes sense, as we belong to the animal world, after all). In essence, it says that we humans implicitly recognize animals—but not plants—as entities related to our own kind. "Plant blindness," however, is more than a consequence of zoocentrism; the concept implies that even when we recognize plants as a form of life, to us, they represent "the other," a fundamentally different kind of entity, one that, although recognized as life, is perceived as unmistakably alien.

Other schools of thought, although agreeing that plant blindness is a real phenomenon, contest the interpretation described above. Their main argument is that this phenomenon is not a matter of willful indifference or a perception of plants as "the other." Rather, these alternate interpretations hold that most humans tend to be unconsciously—some even say *psychologically*—blind to plants as an evolutionary survival mechanism. There are two proposed, yet related interpretations that try to explain this psychological blindness. The first of these is a combination of psychological and computer science concepts that describes how the normal human visual system captures and processes information. In a nutshell, human eyes are normally able to capture about ten million bits of data every second. (A "bit," defined in terms of computer science, is the smallest possible unit of data; essentially, the zeros or ones in a binary system. It is one or the other—there are no intermediate stages; hence the smallest possible data units. This concept goes back to the founder of information theory, Claude E. Shannon. Please see scientificamerican.com/article/claude-e-shannon-founder/.) Of that amount, our brain processes and uses about all of 0.0004 percent (this corresponds to forty bits per second) and eventually our full consciousness is only capable of dealing directly with about sixteen bits per second, a mere 0.00016 percent of the raw data collected by human vision. There is every reason to believe that this applies to other members of the animal kingdom as well. (Please see Allen [2003] and Wandersee and Schussler [1999].) This marked reduction of useful visual data creates a resource allocation problem, which meaningfully limits the data that an animal can use. The obvious solution to this problem is to carefully choose the relevant bits of information—those most likely to enhance the chances of survival. Thus, animals have learned to pay preferential attention to factors like contrast, bright colors, shapes, and perhaps most importantly, movement. These factors give animals the best chance of distinguishing impending danger from the environmental background. This strategy is essentially a solution to the problem of signal-to-noise ratio, because impending danger is almost invariably distinct from the proverbial background noise. In other words, a source of danger that never, ever moves is no dang,er at all. In this context, plants, since they do not usually stand out, are deemed not immediately dangerous and are automatically designated as belonging to the "safe background" category, at least until more information becomes available.

A variant of the psychological blindness interpretation is the "attentional blink" idea. Essentially, attentional blink describes the psychological phenomenon that arises when assessing two unidentified, consecutive stimuli. In general, we see the first stimulus without any problem while the second stimulus goes unnoticed or, alternatively, its identification is impaired or slowed down. It is almost as if the brain assigns the computational resources of the visual system to evaluate only the first stimulus, a type of very short-term habituation, if you will. Attentional blink is a well-established psychophysical phenomenon. (For a brief review of the concept of attentional blink, please see Willems and Martens [2016]. For an exploration of the proposed link between attentional blink and plant blindness, please see Balas and Momsen [2014].)

As in most biological scenarios, the true reason for a particular phenomenon will likely not be simple. Plant blindness will probably not be exclusively either a result of the psychological/data processing or the attentional blink concepts—I am betting the reality will tend to be a combination of both. At any rate, it is a fascinating phenomenon, and one we clearly need more research to completely explain.

3    Please see https://www.npr.org/2020/11/27/938878618/.

4    If you are interested in more examples of animal and microorganism survival strategies, you might want to check out my previous book: *Strange Survivors: How Organisms Attack and Defend in the Game of Life*. BenBella Books (2018).

5    Blue whales and redwood trees are examples of the biggest *individual* organisms. The biggest organism on record is a type of fungus (*Armillaria ostoyae*), a colony actually, that scientists found in Oregon. It covers an area of 2,385 acres. There is some controversy about whether this is an individual organism, but we'll not get into that discussion here. It is undeniable that, whether an individual or a colony, we are talking about a lot of living matter here! Please see scientificamerican.com/article/strange-but-true-largest-organism-is-fungus/. Incidentally, if you wish to know more about some other "extremes" of the living world, you might want to read Matthew La Plante's excellent book, *Superlative* (2019).

6    Please see Chamovitz (2012) for a lay-friendly book about this topic. Also, there is a recent book that, although scientifically rigorous, is far from dry: *Thus Spoke the Plant*, by Dr. Monica Gagliano (2018). This book combines a good summary of the science of plant behavior with an almost poetic appreciation for plants and their relationship with us.

7    Mind you, plants use many more senses than the specific examples that we mention here. Please see Chamovitz (2012), Mancuso and Viola (2015), and Walters (2017).

8    Please see Kelz and Mashour (2019).

9    Please see Baluška and collaborators (2016).

10   Please see Grémiaux and collaborators (2014).

11   Please see Yokawa and collaborators (2018, 2019).

12   In 2014, I said the following about this new field of plant neurobiology:

"Yes, this is an actual field of study, albeit one with a tortuous conception, particularly painful birth, and a very difficult childhood . . . It has not reached its adolescence yet." (In fact, at the time of this writing, plant neurobiology is now a proper teenager. Please see Pagán [2014] *The First Brain: The Neuroscience of Planarians*. Oxford University Press, p. 59.)

Alas, there was no universal welcome for "plant neurobiology" from the general scientific community. A widespread misconception about scientists is that we are dispassionate observers of nature. Well, I have some news for you. We are not dispassionate about nature (or about anything else for that matter); nope, not even close. To be completely fair, it is true that properly conducted science follows a series of loosely organized guidelines that we collectively call the *scientific method*. It is also true that the scientific method is not perfect—far from it. Even so, the scientific method is the best we have so far and has indeed worked pretty well for a couple of hundred years now. By following these guidelines, we have come a long way as far as our knowledge of the natural world is concerned, and are still going strong. But it is also true that science is done by people, and people will be people, in the sense that we can stubbornly hold on—logically or otherwise—to preconceived notions. Furthermore, academics have a propensity toward being quite argumentative over topics of study, about the interpretation of such studies, over the meaning and applicability of a mere word, and so on. I guess that what I am trying to say is that we scientists are not perfect (there's a news flash for you!), but paradoxically, sometimes these very imperfections help advance science. Science progresses not as much by the harmonious consensus of its practitioners, but rather by (oftentimes strong) disagreements. These disagreements are sometimes politely debated, at other times discussed standoffishly, and in many instances argued quite acrimoniously.

I am not going to spend too much time on the "origin story" history of plant neurobiology. (For accessible reviews on how this field came to be, please see Baluška and Mancuso [2005, 2009a, b, c [2013]; Brenner and collaborators [2006]; Iriti [2013]; and Stahlberg [2006].) However, one way of defining the main objective of the field is "to study how plants capture environmental signals and process such signals—via electrochemical impulses—in a way that helps the plant maximize its chances of survival." This sounds reasonable enough, right? The fact is that even though plants do not have brains, nerves, neurons, or anything of the sort, they make good use of electrophysiological and chemical signaling that closely parallels the biology of nerve cells. Furthermore, the original discovery and description of many phenomena related to neurobiology happened using plants, not animals. No scientist disputes this fact, and this is not the source of the controversy about the idea of the neuroscience of plants. The issue is that the scientists who initially championed this idea chose certain terms derived from classical, animal-based neurobiology as metaphors; no more, no less. Despite every disclaimer that plant neurobiologists offered in this regard, some animal neurobiologists took a, let's say, emphatic exception to the use of the "neuro" prefix when talking about plants, and to be fair, it is not hard to understand why. After all, "neurobiology" and "behavior" are words that are overwhelmingly associated with animal life. Therefore, it is hardly surprising that traditional neurobiologists got even more than simply irritated when plant neurobiology advocates began to use terms like "intelligence," "mind," "memory," "cognition," and (gasp!) "brain," applied to plant life. The delicious irony is that these words have no clear, universally accepted connotation even in "animal-centric" circles (as an example, you should hear the ardent debates about what exactly constitutes a brain!) and therefore their exact meaning and definition are still actively debated with no clear, unambiguous consensus in sight. Nonetheless, again, scientists tend to be territorial, and the emergence of a new field or even new points of view applied to an established field often elicit emotional reactions. Human emotions, though ostensibly not actually running the show, do strongly influence scientific debate. (We are not Vulcan, after all.)

13    The arguments that the phrase "plant neurobiology" helped initiate are still going strong, and just looking at the titles of some of the opinion papers on the topic will give you an idea of the civility—or lack thereof—of the discussion. (Please see Alpi and collaborators [2007]; Brenner [2007]; Calvo and collaborators [2019]; Gagliano [2013, 2017, 2018]; Gagliano and collaborators [2014, 2016, 2017, 2018]; and Rehm and Gradmann [2010].) These are interesting times indeed to "think about plant thoughts." If you want to get a more complete picture of how "hot"—and I mostly mean contentious—the area is right now, please see Calvo and collaborators (2017, 2020); Taiz and collaborators (2019, 2020); and Trewavas and collaborators (2020). I would like to especially direct your attention to a paper whose title summarizes my feelings about this issue: "Experiment Rather Than Define." (From Maher [2019].) Plants are clearly more than "second-class living beings" or mere footnotes in the book of life, and I think that at the very least they deserve their own formal set of principles to account for their special kind of behavior.

14    Again, please see Chamovitz (2012) *What a Plant Knows: A Field Guide to the Senses.*

15    For an excellent overview on how plants perceive the world, please see Chamovitz (2012). For another good resource—albeit a shorter and slightly more opinionated one—please see Mancuso and Viola (2015) *Brilliant Green: The Surprising History and Science of Plant Intelligence.*

16    Translated from the original German by Dr. Thomas W. Baumann. Please see Baumann (2006).

17    If you want to know more about venoms in nature, you might want to check out the following popular science books: Wilcox (2016) *Venomous: How Earth's Deadliest Creatures Mastered Biochemistry* and my own, Pagán (2018) *Strange Survivors: How Organisms Attack and Defend in the Game of Life.*

18    Please see insider.si.edu/2015/06/how-carnivorous-plants-avoid-eating-their-pollinating-insect-friends/.

19    Please see Blackwell (2011).

20    Please see Bohacek and collaborators (1996).

21    Upon reading this book, I was sad to learn that the author passed a few years back. I would have loved to have corresponded or had a conversation or two with Dr. Firn. Please see Firn (2009), chapter five. Other less-favored hypotheses in this regard include the idea that these products are simply randomly produced, which is not consistent with what we observe in nature (evolution and all that).

22    Please see Willis (2002).

23    Please see Inderjit and Duke (2003), Arimura and collaborators (2009), Farooq and collaborators (2011), Leão and collaborators (2009), Porter and Targett (1988).

24    Three excellent review articles that discuss allelopathy in more detail are Duke and collaborators (2013), Inderjit and Duke (2003), and Inderjit and collaborators (2011).

25    For a more detailed discussion of subtractive/additive allelopathic interactions between plants, please see Harper (1975).

26    Please see Bennett and Inamdar (2015), Piechulla and collaborators (2015), and Vergara-Fernández and collaborators (2018).

27    For a very good survey of the classes of molecules that can be HIPVs/VOCs, please see Aljbory and Chen (2018).

28    Please see Dicke (2009), Kalske and collaborators (2019), and War and collaborators.

29    In many cases because of evolutionary mechanisms that are discussed elsewhere; please see Bawa (2016), Dudley (2015), and Sachs and collaborators (2004). Also see Pagán (2018), chapter six.

30    Please see Kessler and Baldwin (2002).

31    But you would never know it from the title of the "original" paper: "Ecology of Infochemical Use by Natural Enemies in a Tritrophic Context." Please see Vet and Dicke (1992).

32    Please see Clavijo McCormick and collaborators (2012).

33    Please see Abbas and collaborators (2017), Degenhardt and collaborators (2003), and Peñaflor and Bento (2013).

34    We will not explore evolutionary arms races extensively in this book, as this concept is explained in some detail elsewhere. If you want to know more about biological arms races, I wrote a more extensive introduction in *Strange Survivors*, pp. 13–20. For a more technical exposition of the biological arms races, please see Dawkins and Krebs (1979). Yes, that Dawkins: Richard.

35    For a reader-friendly exploration of this idea, please see Ehrlich and Raven (1967); for a slightly more technical exploration of said topic, please see Ehrlich and Raven (1964).

36    Please see Reynolds and collaborators (2018).

37    Please see bigthink.com/surprising-science/how-magic-mushrooms-evolved.

38    For an extended rendering of this story, please see Boyce (2019): "Psychoactive plant- and mushroom-associated alkaloids from two behavior modifying cicada pathogens."

39    Please see smithsonianmag.com/science-nature/do-insects-have-consciousness-180959484.

40    The many ways that plants (and fungi) have to mess up nervous systems is a truly fascinating topic. If you want to know more about this in a more technical way, I refer you to one of the most complete books on the subject: *Plants and the Human Brain* by Dr. David O. Kennedy.

41    For a concise yet insightful perspective on the differences between foods, drugs, and toxins, please see Raubenheimer and Simpson (2009).

42    Please see Nesse and Berridge (1997), Hagen and collaborators (2009, 2013), Sullivan and collaborators (2008), and Williams and Nesse (1991).

43    Please see Stephens and Dudley (2004).

44    There is a very reader-friendly resource about his idea, *The Drunken Monkey: Why We Drink and Abuse Alcohol*. Please see Dudley (2014), prologue. For the record, I think that this was a very brave rendition of his personal experiences. These are very painful memories indeed.

45    Please see Dudley (2000, 2002, 2014) and Stephens and Dudley (2004).

46    Please see Carrigan and collaborators (2014).

47    We should keep in mind that the capacity to synthesize vitamin C is present in many types of organisms, including primates. However, among mammals, anthropoids (chimpanzees, bonobos, gorillas, orangutans, humans, etc.) alongside (strangely enough) guinea pigs and certain bat species lack the ability to produce vitamin C. Please see Drouin and collaborators (2011).

48    Please see Dudley (2014).

49    Please see Milton (2014) and Wrangham and Wilson (2015).

50    Please see Hagen and collaborators (2013), and Saniotis (2010).

51    For a thorough exploration of this topic, please see Hagen and collaborators (2009, 2013) Sullivan and Hagen (2002), and Sullivan and collaborators (2008).

52    Again, please see Firn (2010), chapter two. Also, please see Evans (2009) for a thorough review of alkaloids.

53    And in case you are interested, here is a short article of mine on alkaloids: decodedscience.org/drugs-nature-chemistry-addictive-alkaloids/53057.

54    Nicotine's role in nature seems to be as a defensive molecule against insects; one of the original pesticides, if you will. Please see Soloway (1976) and Steppuhn and collaborators (2004).

55    For a concise review of neurotransmission, please see Pagán (2014), pp. 41–90.

56    Acetylcholine is probably the best-known neurotransmitter. It is an integral part of the nervous system of all animals. In humans, its deficiency can lead to a variety of conditions including Alzheimer's disease, among others. This neurotransmitter has a fascinating history in terms of how scientists discovered the phenomenon of neurotransmission. In fact, acetylcholine was the very first neurotransmitter ever discovered. For a fascinating review of this history, please see Karczmar (1993).

57    This, in fact, will apply to pretty much any bioactive molecule, in various degrees of severity.

## Chapter Five

1    In addition to ethanol, they tested imipramine and isocarboxazide (antidepressants), pentobarbital (a barbiturate), chlorpromazine (an antipsychotic), d-amphetamine sulphate, lysergic acid diethylamide (LSD), meprobamate and chlordiazepoxide (antianxiety medications), and nialamide (an antidepressant no longer in clinical use).

2    Floru and collaborators (1969).

3    This brief history of killer bees was adapted and expanded from my 2018 book, *Strange Survivors* (Pagán 2018), page 167. For a really nice website with relevant information about this kind of bee, please see propacificbee.com/infographic/AHB/infographic.php.

4    And you may want to check out their paper about it; please see Abramson and collaborators (2002, 2003).

5    Abramson and collaborators (2000).

6    Please see Schmidt (2016) and scienceblogs.com/retrospectacle/2007/05/16/schmidt-pain-index-which-sting. The text in this box was adapted from my review of Dr. Schmidt's book published in baldscientist.com/2016/07/02/the-sting-of-the-wild-by-j-o-schmidt-book-review.

7    Please see Schmidt (2018). In 2019, Dr. Schmidt published an updated review of his scale (Schmidt, 2019).

8    Please see Bosmia and collaborators (2015), del Toro and collaborators (2012), and Dossey (2010).

9    And Dr. Schmidt as well; he is only too aware of the dangers involved. Please see Schmidt (2018).

10    He obtained his PhD from Cornell University's Department of Neurobiology and Behavior and is now at the Max Planck Institute of Ornithology / University of Konstanz, Germany.

11    Please see Smith (2014) "Honeybee Sting Pain Index by Body Location." Full disclosure: I'm a product of Cornell University as well.

12    Please see Barron and Plath (2017), Müller (2018), and Nürnberger and collaborators (2017, 2019). Also, for an especially readable review paper, please see Grüter and Farina (2009).

13    Please see www.theguardian.com/environment/2017/aug/06/country-diary-drunk-bees-incapable-of-flying-1917.

14    Please see theguardian.com/science/2001/dec/13/research.highereducation1.

15    Please see www.uq.edu.au/news/article/2018/06/love-inspires-new-species-name.

16    Please see Bozic and collaborators (2007).

17    Please see Barron and collaborators (2009) and Søvik and collaborators (2014, 2014, 2018).

18    Please see Si and collaborators (2005).

19    Please see Mustard and collaborators (2012).

20    Please see the aptly titled "Unexpectedly Strong Effect of Caffeine on the Vitality of Western Honeybees (*Apis mellifera*)," by Strachecka and collaborators (2014).

21    For a couple of representative reviews on the possible link between caffeine intake and longevity, please see Carman and collaborators (2014) and Bhatti and collaborators (2013).

22    For a brief exploration of why this is so, please see my *Strange Survivors* (2018).

23    For a very reader-friendly exploration of the reproductive activities of bedbugs, please see Pfiester and collaborators (2009). For more technical articles, please see Horgan and collaborators (2011) and Michels and collaborators (2015).

24    Please see Zars (2012).

25    This is part of a general evolutionary mechanism called *sexual selection*. If you want to know more about some aspects of this mechanism, please see Clutton-Brock (2007, 2017) and Stockley and Bro-Jørgensen (2011).

26    If you want to know more about fruit fly sexual behavior (and why wouldn't you?), please see Aranha and Vasconcelos (2018), Villella and Hall (2008), and Yamamoto and collaborators (2014).

27    Please see Devineni Ulrike Heberlein (2012).

28    A peptide is essentially a short protein. Both NPY and NPF come in various "versions" depending on the organism and on certain particularities of how genes are expressed. Please see Nässel and Wegener (2011).

29    For a more extensive account of this work, please see Shohat-Ophir and collaborators (2012). For a really good, reader-friendly summary of this research, please see Zars (2012). This latter review is amusingly titled "She Said No, Pass Me a Beer."

30    Some of these headlines include "Mending a Broken Heart with Alcohol," "Sexually Deprived *Drosophila* Become Bar Flies," "A Fruit Fly Walks into a Bar . . . ," and "Sexually Frustrated Flies Get Drunk," among others. For a more complete list, please see the reference section of Guevara-Fiore and Endler (2014). In our era of "clickbait," these headlines are attention-grabbing indeed.

31    Again, Guevara-Fiore and Endler (2014) provided an objective and fair critique on this matter.

32    For an objective analysis of these ideas, please see Guevara-Fiore and Endler (2014).

33    Zer-Krispil and collaborators (2018) "Ejaculation Induced by the Activation of Crz Neurons Is Rewarding to *Drosophila* Males."

34    If you want to know more about reward mechanisms in insects, particularly *Drosophila*, please see Kaun and collaborators (2012), Lowenstein and Velazquez-Ulloa (2018), and Perry and Barron (2013).

35    For a few good general reviews about corazonin and corazonin-related peptides, please see Boerjan and collaborators (2010), Tsai (2018), and Zandawala and collaborators (2018).

36    Please see Lee and collaborators (2008).

37    The link between dopamine and intermale courtship was further confirmed by subsequent research by Dr. Han's group as well as other teams. Please see Aranda and collaborators (2017), Chen and collaborators (2012), and Liu and collaborators (2009).

38    Please see nature.com/news/2008/080103/full/news.2007.402.html.

39    Please see David and collaborators (1983), Mercot and collaborators (1994), Ogueta and collaborators (2010), and Park and collaborators (2017).

40    Please see Fry (2014).

41    Please see Chakraborty and Fry (2016) and Fry and Saweikis (2006).

42    Please see Kacsoh and collaborators (2013).

43    Please see Bhadra and collaborators (2017), Dominoni and collaborators (2017), Harding (2000), Kuhlman and collaborators (2017), and Dunlap and Loros (2016, 2017).

44    Please see Helfrich-Förster (2005, 2019).

45    Please see Michel and Lyons (2014).

46    For more information on chronotherapy and chronopharmacology, please see Dallmann and collaborators (2016), Ohdo and collaborators (2019), and Tahara and Shibata (2013).

47    These circadian activities have been studied at the molecular level. Please see De Nobrega and Lyons (2016).

48    Please see De Nobrega and Lyons (2017).

49    Please see De Nobrega and collaborators (2017), and De Nobrega and Lyons (2018).

50    For three excellent general books on chronobiology, please see Foster and Kreitzman (2017), Palmer (2002), and an oldie but a goodie: Ward (1971).

51    Later on, Dr. Witt's group and others tested the effect of drugs on other spider species, with similar results. Please see Hesselberg and Vollrath (2004) and Samu and Vollrath (1992).

52    For a very readable introduction to spiders and how they make silk, please see *Spider Silk: Evolution and 400 Million Years of Spinning, Waiting, Snagging, and Mating* (Brunetta and Craig [2012]). This is a fascinating and very beautiful book.

53    From Witt (1954).

54 Incidentally, after his initial experiments, Dr. Witt worked mainly with another species of orb-weaving spider, *Araneus diadematus*, obtaining similar results.

55 Yes, it is true! The spider in question (*Bagheera kiplingi*) belongs to the general class of jumping spiders and, strictly speaking, it is mostly vegetarian. They are known to hunt other animals in a pinch, however. Please see Jackson (2009).

56 Please see Reed and collaborators (1981) and Witt (1971).

57 Please see NASA Tech Briefs, April 1995, page 82.

58 Please see Kandel (1989), Sattelle and Buckingham (2006), and Sweatt (2016). I once met Dr. Kandel at a neuroscience meeting, which triggered a fanboy moment on my part. If you are interested in reading this story, please see baldscientist.com/2014/09/04/and-oldie-the-day-i-met-a-nobel-laureate/.

59 Please see Lee and collaborators, "An Argument for Amphetamine-Induced Hallucinations in an Invertebrate" (2018).

60 Full disclosure: besides being an unapologetic fan, I know a thing or two about the planarian nervous system. I actually wrote the only available book so far dealing with aspects of the pharmacology and neurobiology of planarians. Please see Pagán (2014) *The First Brain: The Neuroscience of Planarians*.

61 If you want to know more about regeneration, please see Pagán (2018) *Strange Survivors: How Organisms Attack and Defend in the Game of Life*, published by the good people of BenBella Books. For more technical explorations of this topic, please see Birkholz and collaborators (2019), Levin and collaborators (2019), Pagán (2014, 2017), and Reddien (2018).

62 Please see Margotta and collaborators (1997) and Palladini and collaborators (1996).

63 I explored the fascinating story of the beginnings of planarian pharmacology in Pagán (2014). Also, please see Raffa and Rawls (2008).

64 Please see Pagán and collaborators (2008, 2009), Raffa and Rawls (2008), Raffa and Desai (2005), and Rowlands and Pagán (2008).

65 To see some of the work generated by my students and myself, please see Bach and collaborators (2016), Baker and collaborators (2011), Pagán (2017, 2019), Pagán and collaborators (2008, 2009, 2012, 2013, 2015), and Schwarz and collaborators (2011).

66 If you want to know more about my research, I wrote a general review article covering it. Please see Pagán (2017) "Planaria: An Animal Model That Integrates Development, Regeneration and Pharmacology," published in the *International Journal of Developmental Biology*.

67 Please see Mohammed and collaborators (2018), Nayak and collaborators (2019), and Tallarida and collaborators (2014).

68 For an excellent popular book on this topic, please see Dr. Robert Sapolsky's *Why Zebras Don't Get Ulcers*, now in its third edition.

69 Please see Deslauriers and collaborators (2018), Flandreau and Toth (2018), and Schöner and collaborators (2017).

70 Please see Zewde and collaborators (2018).

71 Please see Cho and collaborators (2019).

72 For a very readable account of anecdotal experiences describing the cognitive capabilities of octopi as well as the way they express true personalities, please see Montgomery (2016) *The Soul of an Octopus: A Surprising Exploration into the Wonder of Consciousness*. For a more technical exploration of cephalopod cognition, please see Hochner (2012) and Mather and Dickel (2017).

73 There is the possibility that social organisms express their own form of advanced intelligence. I explored this and several other examples of brains and brain-like structures and ensembles in a paper that was part of the proceedings of a working group at the Santa Fe Institute. Please see Pagán (2019) "The Brain: A Concept in Flux."

74 In Dr. Godfrey-Smith's *Other Minds: The Octopus, the Sea, and the Deep Origins of Consciousness* (2016). This is a fascinating exploration of consciousness in general and of octopus consciousness in particular. You will like it.

75 Please see Steele and collaborators (2018) "Cause of Cambrian Explosion—Terrestrial or Cosmic?" This is an intriguing paper, to be sure, that touches upon a few things besides octopi, but I do not buy the

octopus part of it based on what we know about the evolution of the cephalopods. I am not alone in this interpretation.

76    Please see Albertin and collaborators (2015) "The Octopus Genome and the Evolution of Cephalopod Neural and Morphological Novelties." Incidentally, the way this paper is titled is a pet peeve of mine. This paper references *the octopus genome* in its very title, when in reality they sequenced the genome of *one species* of octopus, namely *Octopus bimaculoides*, the protagonist of this story.

77    Please see Edsinger and Dölen (2018).

78    Reviewed in Bershad and collaborators (2016), Betzler and collaborators (2017), and Feduccia and collaborators (2018).

79    For a couple of reviews exploring these intriguing possibilities, please see Schenk and Newcombe (2018) and Sessa and collaborators (2019).

80    Reviewed in Bershad and collaborators (2016), Betzler and collaborators (2017), and Feduccia and collaborators (2018).

## Chapter Six

1    There are anecdotal accounts of reptiles engaging in geophagy (chapter three), but I could not find any specific studies reporting this behavior (please see Jain and collaborators [2008] and Young and collaborators [2011]).

2    Please see Pandey and Verma (2017).

3    For more general information on marine pharmacology, please see Kaul and Daftari (1986), Malve (2016), Mayer and Gustafson (2008), Mayer and collaborators (2017), and Rodríguez (1995).

4    Please see Ciereszko and collaborators (1960), Wahlberg and Eklund (1992), Ferchmin and collaborators (2009), Pagán (1998, 2005, 2014), Pagán and collaborators (2001, 2009), Elkhawas and collaborators (2020), and Yan and collaborators (2019).

5    Please see De Simone and collaborators (2017).

6    Please see Sladky and collaborators (2008).

7    For excellent overviews of environmental toxicology, please see Pesce and collaborators (2018), and Wu and Li (2018).

8    Please see nypost.com/2019/07/15/tennessee-police-advise-public-over-threat-of-meth-gators/.

9    Please see allaboutbirds.org/guide/Cedar_Waxwing/lifehistory.

10    Please see Kinde and collaborators (2011).

11    Please see Eriksson and Nummi (1983), Fitzgerald and collaborators (1990), and Stephen and Walley (2000). In 2012, Duff and collaborators published a similar study.

12    Please see Fossler (1936).

13    For example, please see Levey and Cipollini (1998).

14    Please see Struempf and collaborators (1999).

15    Please see Olson and collaborators (2014).

16    Please see Hyland Bruno and Tchernichovski (2019).

17    Please see Sánchez and collaborators (2004, 2008).

18    Please see Sánchez and collaborators (2010).

19    If you must know, the species in question were *Artibeus jamaicensis, A. lituratus, A. phaeotis, Carollia sowelli, Glossophaga soricina*, and *Sturnira lilium*. Please see Orbach and collaborators (2010).

20    As reported in Morris and collaborators (2006) based on Webb and collaborators (translators, 1990).

21    Please see Morris and collaborators (2006) "Myth, Marula, and Elephant: An Assessment of Voluntary Ethanol Intoxication of the African Elephant (*Loxodonta africana*) Following Feeding on the Fruit of the Marula Tree (*Sclerocarya birrea*)."

22    Please see https://theconversation.com/elephants-get-drunk-because-they-cant-metabolize-alcohol-like-us-137475.

23    Please see Janiak and collaborators (2020) "Genetic evidence of widespread variation in ethanol metabolism among mammals: revisiting the 'myth' of natural intoxication."

24    Please see Siegel and Brodie (1984), and Siegel (2005), chapter five, for a more reader-friendly narrative of the experiments, including some dramatic occurrences not described in the paper itself.

25    A great source of information for all things *Star Trek* is memory-alpha.fandom.com.

26    Please see Greven (2009) "Gender and Sexuality in Star Trek: Allegories of Desire in the Television Series and Films."

27    Please see Chave and collaborators (2019) and Ganswindt and collaborators (2010).

28    From www.nimh.nih.gov/health/topics/schizophrenia/raise/what-is-psychosis.

29    Please see Lindstedt and Nishikawa (2015). The main idea is that, pharmacologically speaking, there are a series of parameters that closely depend on the mass of the animal but are not in direct proportion to the weight of the animal and rather have to be calculated using power laws. These factors include drug metabolism, drug delivery, and elimination, among other factors. For a discussion of the complexity of scale laws, please see Dokoumetzidis and Macheras (2009), Marquet and collaborators (2005), and Nevill and collaborators (2004).

30    Please see West and Collaborators (1962).

31    Please see Ji and collaborators (2009).

32    Please see Sparks and collaborators (2017).

33    Please see Nestorović and collaborators (2010).

34    Please see Aydin and collaborators (1998). If we think about it, this is not so surprising after all. In its unprocessed form, catnip has a smell strongly reminiscent of mint, and mint can act as a weak local analgesic.

35    Please see Espín-Iturbe and collaborators (2017).

36    Please see Bol and collaborators (2017).

37    Please see Siegel (2005), page 62.

38    Please see Birkett and Pickett (2003) and Eisner (1964).

39    Please see Bol and collaborators (2017).

40    Please see Bol and collaborators (2017).

41    From Siegel (2005), page 62.

42    Please see Wells and Egli (2004). Despite its small size (no bigger than house cats, actually slightly smaller on average), the black-footed cat is probably one of the most interesting wild felines. It is a highly efficient hunter, putting to shame its much bigger cousins (please see livescience.com/63992-deadliest-cat.html).

43    For thorough explorations of DMT/ayahuasca, please see McKenna (1992, 1996). We met Dr. McKenna when discussing his stoned ape hypothesis in chapter two.

44    Please see singingtotheplants.com/2009/02/jaguar-on-ayahuasca/.

45    According to Siegel (2005), some of these animals included "sheep, antelopes, pigs, rabbits, hens, bees, and even an insect grub or two."

46    These plants have a worldwide distribution, but the earliest descriptions are from North America, where their economic impact is best documented (Wu and collaborators [2014, 2016]).

47    As reported by Bender (1983).

48    Please see Colegate and collaborators (1979) and Molyneux and James (1982).

49    I know, fungi again. The technical term is endophytic fungi. For more information, please go to Nzabanita and collaborators (2018) and Moore and Johnson (2017).

50    Please see Siegel (2005), pages 50–53.

51    Please see theguardian.com/world/2009/jun/25/wallabies-high-tasmania-poppy-fields.

52    Please see http://news.bbc.co.uk/2/hi/asia-pacific/8118257.stm.

53    Please see www.al.com/news/2019/06/attack-squirrel-rescued-at-scene-of-alabama-meth-bust.html.

54    Please see Crocq (2007) and Siegel (2005), pages 65–70.

55    Please see Siegel (2005), page 65.

56    Please see Arthur (2003), van Renterghem (1995), and Goldhor (2012) in namyco.org/docs/MycophileNovDec2012.pdf. Also, herbarium.0-700.pl/biblioteka/Mushrooms%20and%20Mankind.pdf, livescience.com/25731-magic-mushrooms-santa-claus.html, livescience.com/42077-8-ways-mushrooms-explain-santa.html, realitysandwich.com/shaman-claus-the-shamanic-origins-of-christmas/, livescience.com/16286-hallucinogens-lsd-mushrooms-ecstasy-history.html.

57    Please see Bowler (2005) and patheos.com/blogs/panmankey/2018/12/santa-was-not-a-shaman/.

58    For an excellent overview of the parallels between dolphins and primates, please see Bearzi and Stanford (2008) *Beautiful Minds*.

59    Please see Bearzi and Stanford (2008) and Marino (2002).

60    Please see King and collaborators (2019).

61    Please see Gregg (2013), chapter six, and Patterson and collaborators (1998).

62    Please see Tamaki and collaborators (2006).

63    Please see Cocconi and Morrison (1959).

64    This a fascinating topic that is nonetheless very far from what we want to talk about in this book. I would like to refer you to excellent resources. Please see Bizoni (2012), Cocconi and Morrison (1959), Kellermann and collaborators (2020), and Sagan (2000).

65    Please see Oberhaus (2019), page 39.

66    Please see Lilly and Miller (1961).

67    Please see Sagan (2000), page 170.

68    Please see Clarke (2014).

69    Please see Clarke (2014).

70    Please see https://www.samorini.it/doc1/alt_aut/ad/abramson-the-use-of-lsd-in-psychotherapy-and-alcoholism.pdf.

71    Steiner (1995) "Rough-Toothed Dolphin, S*teno bredanensis*: A New Species Record for the Azores, with Some Notes on Behavior."

72    Please see Barber (2016).

73    Please see Kuczaj and Eskelinen (2014) and Kuczaj and Yeater (2007).

74    *Spy in the Pod*, an original BBC series broadcast in 2014 and later released on the US-based Discovery Channel. An innovative aspect of this series was their use of automated underwater cameras disguised as various objects/organisms that would not be out of place in the ocean (various fish species and even a mock-up dolphin). In this way, the filmmakers hoped to catch the behavior of dolphins, as "naturally" as possible. Please see bbc.co.uk/programmes/b03ncs0h and pbs.org/wnet/nature/spy-in-the-pod-about/15270/.

75    Please see dailymail.co.uk/sciencetech/article-2530664/High-not-dry-Dolphins-filmed-chewing-toxic-puffer-fish-enjoy-narcotic-like-effects.html.

76    Please see smithsonianmag.com/smart-news/dolphins-seem-to-use-toxic-pufferfish-to-get-high-180948219/.

77    I talked at some length about tetrodotoxin in *Strange Survivors* (Pagán, 2018), in a section titled "A Rather Wicked Molecule" (pages 15–20), and its ecological significance in the relationship between a certain species of newt (*Taricha granulosa*, among others) and garter snakes (*Thamnophis sirtalis* and related species; pages 18–20). Please also see Bane and collaborators (2014), Brodie (2009), Hanifin (2010), Lago and collaborators (2015), Narahashi (2008), Ritson-Williams and collaborators (2006), and Yamada and collaborators (2017).

78    Incidentally, there is a hypothetical link between tetrodotoxin and zombie lore (please see Pagán, 2018, page 17).

79    Please see Hokama and collaborators (1990).

80    She is an award-winning blogger and science writer with a doctorate in Cell and Molecular Biology with a specialization in Ecology, Evolution, and Conservation Biology (from christiewilcox.com).

81    Please see Wilcox (2016) *Venomous*.

82    Please see https://www.discovermagazine.com/planet-earth/do-stoned-dolphins-give-puff-puff-pass-a-whole-new-meaning.

83    Please see Hwang and Noguchi (2007), Pagán (2018), Pratheepa and Vasconcelos (2017), and Wu and collaborators (2005).

84    Please see Dawkins (2016) *The Selfish Gene: 40th Anniversary Edition*. Oxford University Press.

# BIBLIOGRAPHY AND FURTHER READING

Abbas, F. et al. 2017. Volatile terpenoids: multiple functions, biosynthesis, modulation and manipulation by genetic engineering. *Planta* 246 (5): 803–816.

Abbott, J. 2014. Self-medication in insects: current evidence and future Perspectives. *Ecological Entomology* 39: 273–280.

Abramson, C. I. and I. S. Aquino. 2002. Behavioral studies of learning in the Africanized honey bee (*Apis mellifera L.*). *Brain Behav Evol* 59 (1–2): 68–86.

Abramson, C. I., G. W. Fellows, B. L. Browne, A. Lawson, and R. A. Ortiz. 2003. Development of an ethanol model using social insects: II. Effect of Antabuse on consumatory responses and learned behavior of the honey bee (*Apis mellifera L.*). *Psychol Rep* 92 (2): 365–78.

Abramson, C. I., S. M. Stone, R. A. Ortez, A. Luccardi, K. L. Vann, K. D. Hanig, and J. Rice. 2000. The development of an ethanol model using social insects I: behavior studies of the honey bee (*Apis mellifera L.*). *Alcohol Clin Exp Res* 24 (8): 1153–66.

Adolphs, R., D. Anderson. 2013. Social and emotional neuroscience. *Curr Opin Neurobiol* 23 (3): 291–3.

Akhavan, B. J., L. Ostrosky-Zeichner, and E. J. Thomas. 2019. Drunk Without Drinking: A Case of Auto-Brewery Syndrome. *ACG Case Rep J* 6 (9): e00208.

Agrawal, A. A. 2017. *Monarchs and Milkweed: A Migrating Butterfly, a Poisonous Plant, and Their Remarkable Story of Coevolution*. Princeton, NJ: Princeton University Press.

Agrawal, A. A., A. P. Hastings. 2019. Trade-offs constrain the evolution of an inducible defense within but not between plant species. *Ecology*: e02857.

Ahmed, S. H. 2012. The science of making drug-addicted animals. *Neuroscience* 211: 107–25.

Albergaria, H., and N. Arneborg. 2016. Dominance of *Saccharomyces cerevisiae* in alcoholic fermentation processes: role of physiological fitness and microbial interactions. *Appl Microbiol Biotechnol* 100 (5): 2035–46.

Albertin, C. B., O. Simakov, T. Mitros, Z. Y. Wang, J. R. Pungor, E. Edsinger-Gonzales, S. Brenner, C. W. Ragsdale, and D. S. Rokhsar. 2015. The octopus genome and the evolution of cephalopod neural and morphological novelties. *Nature* 524 (7564): 220–4.

Alexander, S. and C. de B. Webb, eds. 1990. *Adulphe Delegorgue's Travels in Southern Africa*. Translated by F. Webb. Vol. 1. Pietermaritzburg, South Africa: University of KwaZulu-Natal Press.

Allen, W. 2003. Plant Blindness. *BioScience* 53 (10): 96.

Aljbory, Z., M. S. Chen. 2018. Indirect plant defense against insect herbivores: a review. *Insect Sci* 25 (1): 2–23.

Alpi, A. et al. 2007. Plant neurobiology: no brain, no gain? *Trends Plant Sci* 12 (4): 135–6.

Alvarez, S., P. Marcasuzaa, and L. Billon. 2021. Bio-Inspired Silica Films Combining Block Copolymers Self-Assembly and Soft Chemistry: Paving the Way toward Artificial Exosqueleton of Seawater Diatoms. *Macromol Rapid Commun* 42 (4): e2000582.

Anderson, D. J., R. Adolphs. 2014. A framework for studying emotions across species. *Cell* 157 (1): 187–200.

Anthony, W. E. et al. 2015. Testing Dose-Dependent Effects of the Nectar Alkaloid Anabasine on Trypanosome Parasite Loads in Adult Bumble Bees. *PLoS One* 10 (11): e0142496.

Antolin, M. F. 2011. Evolution, Medicine, and the Darwin Family. *Evo Edu Outreach* (2011) 4: 613–623.

Aranda, G. P., S. J. Hinojos, P. R. Sabandal, P. D. Evans, and K. A. Han. 2017. Behavioral Sensitization to the Disinhibition Effect of Ethanol Requires the Dopamine/Ecdysone Receptor. *Front Syst Neurosci* 11: 56.

Aranha, M. M., and M. L. Vasconcelos. 2018. Deciphering *Drosophila* female innate behaviors. *Curr Opin Neurobiol* 52: 139–148.

Arimura, G., K. Matsui, J. Takabayashi. 2009. Chemical and molecular ecology of herbivore-induced plant volatiles: proximate factors and their ultimate functions. *Plant Cell Physiol* 50 (5): 911–23.

Arthur, J. 2003. *Mushrooms and Mankind: The Impact of Mushrooms on Human Consciousness and Religion*. Book Tree.

Aydin, S., R. Beis, Y. Oztürk, and K. H. Baser. 1998. Nepetalactone: a new opioid analgesic from *Nepeta caesarea* Boiss. *J Pharm Pharmacol* 50 (7): 813–7.

Auesukaree, C. 2017. Molecular mechanisms of the yeast adaptive response and tolerance to stresses encountered during ethanol fermentation. *J Biosci Bioeng* 124 (2): 133–142.

Bach, D. J., M. Tenaglia, D. L. Baker, S. Deats, E. Montgomery, and O. R. Pagán. 2016. Cotinine antagonizes the behavioral effects of nicotine exposure in the planarian *Girardia tigrina*. *Neurosci Lett* 632: 204–8.

Baker, D., S. Deats, P. Boor, J. Pruitt, and O. R. Pagán. 2011. Minimal structural requirements of alkyl γ-lactones capable of antagonizing the cocaine-induced motility decrease in planarians. *Pharmacol Biochem Behav* 100 (1): 174–9.

Balas, B., and J. L. Momsen. 2014. Attention blinks differently for plants and animals. *CBE Life Sci Educ* 13 (3): 437–43.

Baluška, F., S. Mancuso. 2007. Plant neurobiology as a paradigm shift not only in the plant sciences. *Plant Signal Behav* 2 (4): 205–7.

Baluška, F. 2009a. Deep evolutionary origins of neurobiology: Turning the essence of 'neural' upside-down. *Commun Integr Biol* 2 (1): 60–5.

Baluška, F. 2009b. Plant neurobiology: from sensory biology, via plant communication, to social plant behavior. *Cogn Process* 10 Suppl 1: S3–7.

Baluška, F. 2009c. Plant neurobiology: From stimulus perception to adaptive behavior of plants, via integrated chemical and electrical signaling. *Plant Signal Behav* 4 (6): 475–6.

Baluška, F. et al. 2005. Plant synapses: actin-based domains for cell-to-cell communication. *Trends Plant Sci* 10 (3): 106–11.

Baluška, F. et al. 2009. The 'root-brain' hypothesis of Charles and Francis Darwin: Revival after more than 125 years. *Plant Signal Behav* 4 (12): 1121–7.

Baluška, F. et al. 2016. Understanding of anesthesia—Why consciousness is essential for life and not based on genes. *Commun Integr Biol* 9 (6): e1238118.

Baluška, F., S. Mancuso. 2013. Ion channels in plants: from bioelectricity, via signaling, to behavioral actions. *Plant Signal Behav* 8 (1): e23009.

Bane, V., M. Lehane, M. Dikshit, A. O'Riordan, A. Furey. 2014. Tetrodotoxin: Chemistry, toxicity, source, distribution and detection. Toxins (Basel) 6: 693–755.

Baracchi, D., M. Lihoreau, and M. Giurfa. 2017. Do Insects Have Emotions? Some Insights from Bumble Bees. *Front Behav Neurosci* 11: 157.

Baracchi, D. et al. 2015. Behavioural evidence for self-medication in bumblebees? *F1000Res* 4: 73.

Barber, T. M. 2016. Preliminary Report of Object Carrying Behavior by Provisioned Wild Australian Humpback Dolphins (*Sousa sahulensis*) in Tin Can Bay, Queensland, Australia. *International Journal of Comparative Psychology* 29 (1).

Barron, A. B., R. Maleszka, P. G. Helliwell, and G. E. Robinson. 2009. Effects of cocaine on honey bee dance behaviour. *J Exp Biol* 212 (Pt 2): 163–8.

Barron, A. B. and J. A. Plath. 2017. The evolution of honey bee dance communication: a mechanistic perspective. *J Exp Biol* 220 (Pt 23): 4339–4346.

Barrows, W. M. 1915. The reactions of an orb-weaving spider, *Epeira scloperetaria* (Cl.) to rhythmic vibrations of its net. *Biol Bull* 29: 316–332.

Basu, N. 2015. Applications and implications of neurochemical biomarkers in environmental toxicology. *Environ Toxicol Chem* 34 (1): 22–9.

Bateson, M. et al. 2011. Agitated honeybees exhibit pessimistic cognitive biases. *Curr Biol* 21 (12): 1070–3.

Baumann, T. W. 2006. Some thoughts on the physiology of caffeine in coffee—and a glimpse of metabolite profiling. *Braz J Plant Physiol* 18 (1): 243–251.

Bawa, K. S. 2016. Kin selection and the evolution of plant reproductive traits. *Proc Biol Sci* 283 (1842). pii: 20160789.

Bearzi, M., C. Stanford. 2008. *Beautiful Minds: The Parallel Lives of Great Apes and Dolphins*. Harvard University Press.

Belser-Ehrlich, S., A. Harper, J. Hussey, R. Hallock. 2013. Human and cattle ergotism since 1900: symptoms, outbreaks, and regulations. *Toxicol Ind Health* 29 (4): 307–16.

Bender, G. A. 1983. Searching the Southwest for Medicines. *The Journal of Arizona History* 24 (2) 103–118.

Bennett, J. W., A. A. Inamdar. 2015. Are Some Fungal Volatile Organic Compounds (VOCs) Mycotoxins? *Toxins (Basel)* 7 (9): 3785–804.

Benyus, J. 2002. *Biomimicry: Innovation Inspired by Nature*. New York, NY: Harper Perennial.

Berger, K. H., U. Heberlein, and M. S. Moore. 2004. Rapid and chronic: two distinct forms of ethanol tolerance in *Drosophila*. *Alcohol Clin Exp Res* 28 (10): 1469–80.

Berger, K. H., E. C. Kong, J. Dubnau, T. Tully, M. S. Moore, and U. Heberlein. 2008. Ethanol sensitivity and tolerance in long-term memory mutants of *Drosophila melanogaster*. *Alcohol Clin Exp Res* 32 (5): 895–908.

Bershad, A. K., M. A. Miller, M. J. Baggott, and H. de Wit. 2016. The effects of MDMA on socio-emotional processing: Does MDMA differ from other stimulants? *J Psychopharmacol* 30 (12): 1248–1258.

Betzler, F., L. Viohl, and N. Romanczuk-Seiferth. 2017. Decision-making in chronic ecstasy users: a systematic review. *Eur J Neurosci* 45 (1): 34–44.

Bhadane, B. S. et al. 2018. Ethnopharmacology, phytochemistry, and biotechnological advances of family *Apocynaceae*: A review. *Phytother Res* 32 (7): 1181–1210.

Bhadra, U., N. Thakkar, P. Das, M. Pal Bhadra. 2017. Evolution of circadian rhythms: from bacteria to human. *Sleep Med* 35: 49–61.

Bhatti, S. K., J. H. O'Keefe, and C. J. Lavie. 2013. Coffee and tea: perks for health and longevity? *Curr Opin Clin Nutr Metab Care* 16 (6): 688–97.

Birkett, M. A., J. A. Pickett. 2017. Aphid sex pheromones: from discovery to commercial production. *Behav Processes*. 142: 110–115.

Birkholz, T. R., A. V. Van Huizen, and W. S. Beane. 2019. Staying in shape: Planarians as a model for understanding regenerative morphology. *Semin Cell Dev Biol* 87: 105–115.

Bivin, W. S., and B. N. Heinen. 1985. Production of ethanol from infant food formulas by common yeasts. *J Appl Bacteriol* 58 (4): 355–7.

Blackmore, S. 2017. *Consciousness: A Very Short Introduction*. Oxford, UK: Oxford University Press.

Blackwell, M. 2011. The fungi: 1, 2, 3 . . . 5.1 million species? *Am J Bot* 98 (3): 426–38.

Blum, M. S. et al. 1981. Fate of cocaine in the lymantriid *Eloria noyesi*, a predator of *Erythroxylum coca*. *Phytochemistry* 20 (11): 2499–2500.

Boerjan, B., P. Verleyen, J. Huybrechts, L. Schoofs, and A. De Loof. 2010. In search for a common denominator for the diverse functions of arthropod corazonin: a role in the physiology of stress? *Gen Comp Endocrinol* 166 (2): 222–33.

Boese, A. 2007. *Elephants on Acid and Other Bizarre Experiments*. Orlando, FL: Mariner Books.

Bohacek, R. S., C. McMartin, W. C. Guida. 1996. The art and practice of structure-based drug design: a molecular modeling perspective. *Med Res Rev* 16 (1): 3–50.

Bol, S. et al. 2017. Responsiveness of cats (Felidae) to silver vine (*Actinidia polygama*), Tatarian honeysuckle (*Lonicera tatarica*), valerian (*Valeriana officinalis*) and catnip (*Nepeta cataria*). *BMC Vet Res* 13 (1): 70.

Boppré, M. 1984. Redefining pharmacophagy. *J Chem Ecol* 10 (7): 1151–4.

Bosmia, A. N., R. S. Tubbs, C. J. Griessenauer, V. Haddad Jr. 2015. Ritualistic envenomation by bullet ants among the Sateré-Mawé Indians in the Brazilian Amazon. *Wilderness Environ Med* 26 (2): 271–3.

Boucher, D.H. and V. Moeser. 1991. Cocaine and the coca plant. *Bioscience* 41 (2): 72–76.

Bowler, G. 2005. *Santa Claus: A Biography*. McClelland and Stewart.

Bowman, W. C. 2006. Neuromuscular block. *Br J Pharmacol* 147 Suppl 1: S277–86.

Boyce, G. R. et al. 2019. Psychoactive plant- and mushroom-associated alkaloids from two behavior modifying cicada pathogens. *Fungal Ecology* 41: 147e164.

Boys, C. V. 1880. The influence of a Tuning Fork on the Garden Spider. *Nature* 23: 149–150.

Bozic, J., J. DiCesare, H. Wells, C. I. Abramson. 2007. Ethanol levels in honeybee hemolymph resulting from alcohol ingestion. *Alcohol* 41 (4): 281–4.

Braidwood, R. J., J. D. Sauer, H. Helbaek, P. C. Mangelsdorf, H. C. Cutler, C. S. Coon, R. Linton, J. Steward and A. L. Oppenheim. 1953. Did man once live by beer alone? *Am Anthropol* 55 (4): 515–526.

Braitman, L. 2015. *Animal Madness: Inside Their Minds*. New York, NY: Simon & Schuster.

Breivogel, C. S., J. M. McPartland, and B. Parekh. 2018. Investigation of non-CB. *J Recept Signal Transduct Res* 38 (4): 316–326.

Brenner, E. D. et al. 2007. Response to Alpi et al.: plant neurobiology: the gain is more than the name. *Trends Plant Sci* 12 (7): 285–6.

Brodie, E. D. 2009. Toxins and venoms. *Curr Biol* 19 (20): R931–5.

Brøndsted, H. V. 1969. *Planarian Regeneration* (International Series of Monographs in Pure and Applied Biology. Division: Zoology). London, UK: Pergamon Press.

Brook, K., J. Bennett, S. P. Desai. 2017. The Chemical History of Morphine: An 8000-year Journey, from Resin to de-novo Synthesis. *J Anesth Hist* 3 (2): 50–55.

Brookes, M. 2002. *Fly: The Unsung Hero of Twentieth Century Science*. New York, NY: Ecco.

Brower, L. P., S. C. Glazier. 1975. Localization of heart poisons in the monarch butterfly. *Science* 188 (4183): 19–25.

Browne, J. 2008. *Darwin's Origin of Species: Books That Changed the World*. Grove Press.

Brunetta, L. and C. L. Craig. 2012. *Spider Silk: Evolution and 400 Million Years of Spinning, Waiting, Snagging, and Mating*. New Haven, CT: Yale University Press.

Bush, S. E., D. H. Clayton. 2018. Anti-parasite behaviour of birds. *Philos Trans R Soc Lond B Biol Sci* 373 (1751): 20170196.

Buttarelli, F. R., C. Pellicano, and F. E. Pontieri. 2008. Neuropharmacology and behavior in planarians: translations to mammals. *Comp Biochem Physiol C Toxicol Pharmacol* 147 (4): 399–408.

Buttarelli, F. R., F. E. Pontieri, V. Margotta, and G. Palladini. 2002. Cannabinoid-induced stimulation of motor activity in planaria through an opioid receptor-mediated mechanism. *Prog Neuropsychopharmacol Biol Psychiatry* 26 (1): 65–8.

Calabrese, E. J. 2016. The Emergence of the Dose-Response Concept in Biology and Medicine. *Int J Mol Sci* 17 (12).

Calatayud, J., A. González. 2003. History of the development and evolution of local anesthesia since the coca leaf. *Anesthesiology* 98 (6): 1503–8.

Caldwell, R. L., R. Ross, A. Rodaniche, and C. L. Huffard. 2015. Behavior and Body Patterns of the Larger Pacific Striped Octopus. *PLoS One* 10 (8): e0134152.

Calvo, P. et al. 2017. Are plants sentient? *Plant Cell Environ* 40 (11): 2858–2869.

Calvo, P. et al. 2019. Plants are intelligent, here's how. *Ann Bot*; 125 (1): 11–28.

Campbell, N. A., E. Rodríguez. 1996. A Conversation with . . . Eloy Rodríguez. *The American Biology Teacher* 58 (5): 282–286.

Carman, A. J., P. A. Dacks, R. F. Lane, D. W. Shineman, and H. M. Fillit. 2014. Current evidence for the use of coffee and caffeine to prevent age-related cognitive decline and Alzheimer's disease. *J Nutr Health Aging* 18 (4): 383–92.

Carrigan, M. A. et al. 2014 Hominids adapted to metabolize ethanol long before human-directed fermentation. *Proc Natl Acad Sci USA* 112: 458–463.

Carter, K. et al. 2006. Repeated cocaine effects on learning, memory and extinction in the pond snail *Lymnaea stagnalis*. *J Exp Biol* 209 (Pt 21): 4273–82.

Cates, H. M. et al. 2019. National Institute on Drug Abuse genomics consortium white paper: Coordinating efforts between human and animal addiction studies. *Genes Brain Behav* 18 (6): e12577.

Chamovitz, D. 2012. *What a Plant Knows: A Field Guide to the Senses*. Oxford, UK: OneWorld Publications.

Chaboo, C. S. et al. 2016. Beetle and plant arrow poisons of the Juǀʼhoan and Haiǀǀom San peoples of Namibia (*Insecta, Coleoptera, Chrysomelidae; Plantae, Anacardiaceae, Apocynaceae, Burseraceae*). *Zookeys* (558): 9–54.

Chakraborty, M., and J. D. Fry. 2011. *Drosophila* lacking a homologue of mammalian ALDH2 have multiple fitness defects. *Chem Biol Interact* 191 (1–3): 296–302.

Chakraborty, M., and J. D. Fry. 2016. Evidence that Environmental Heterogeneity Maintains a Detoxifying Enzyme Polymorphism in *Drosophila melanogaster*. *Curr Biol* 26 (2): 219–223.

Chave, E., K. L. Edwards, S. Paris, N. Prado, K. A. Morfeld, J. L. Brown. 2019. Variation in metabolic factors and gonadal, pituitary, thyroid, and adrenal hormones in association with musth in African and Asian elephant bulls. *Gen Comp Endocrinol* 276: 1–13.

Chen, A. et al. 2012. Mutation of *Drosophila* dopamine receptor DopR leads to male-male courtship behavior. *Biochem Biophys Res Commun* 423 (3): 557–63.

Chen, A. et al. 2013. Dispensable, redundant, complementary, and cooperative roles of dopamine, octopamine, and serotonin in *Drosophila melanogaster*. *Genetics* 193 (1): 159–76.

Cheng, F., Z. Cheng. 2015. Research Progress on the use of Plant Allelopathy in Agriculture and the Physiological and Ecological Mechanisms of Allelopathy. *Front Plant Sci* 6: 1020.

Chen, R., X. Wu, H. Wei, D. D. Han, H. H. Gu. 2006. Molecular cloning and functional characterization of the dopamine transporter from *Eloria noyesi*, a caterpillar pest of cocaine-rich coca plants. *Gene* 366 (1): 152–60.

Cho, M., S. U. Nayak, T. Jennings, C. S. Tallarida, and S. M. Rawls. 2019. Predator odor produces anxiety-like behavioral phenotype in planarians that is counteracted by fluoxetine. *Physiol Behav* 206: 181–184.

Choisy, M., J. C. de Roode. 2014. The ecology and evolution of animal medication: genetically fixed response versus phenotypic plasticity. *Am Nat* 184 Suppl 1: S31-46.

Clarke, B. 2014. John Lilly, The Mind of the Dolphin, and Communication Out of Bounds. *communication +1*: Vol. 3, Article 8.

Clavijo McCormick, A. et al. 2012. The specificity of herbivore-induced plant volatiles in attracting herbivore enemies. *Trends Plant Sci* 17 (5): 303–10.

Clutton-Brock, T. 2007. Sexual selection in males and females. *Science* 318 (5858): 1882–5.

Clutton-Brock, T. 2017. Reproductive competition and sexual selection. *Philos Trans R Soc Lond B Biol Sci* 372 (1729).

Clutton-Brock, T., S. J. Hodge, G. Spong, A. F. Russell, N. R. Jordan, N. C. Bennett, L. L. Sharpe, and M. B. Manser. 2006. Intrasexual competition and sexual selection in cooperative mammals. *Nature* 444 (7122): 1065–8.

Cocteau, J. 1958. *Opium: The Diary of a Cure*. Translated from the original French by Margaret Crosland and Sinclair Road. Grove Press.

Cobo, B. 1653. *Historia del Nuevo Mundo*. Manuscrito en Lima, Perú, 1653, libro 5°, capítulo XXIX.

Cocconi, G. and P. Morrison. 1959. Searching for Interstellar Communications. *Nature* 184: 844–46.

Colegate, S. M., P. R. Dorling, and C. R. Huxtable. 1979. A spectroscopic investigation of swainsoine: An a-mannosidase inhibitor isolated from *Swainsona canescens*. *Australian Journal of Chemistry* 32: 2257–2264.

Conselice, C. J. et al. 2016. The evolution of galaxy number density at z < 8 and its implications. *The Astrophysical Journal* 830 (83): 1–16.

Cordell, B. J. 2019. *My Gut Makes Alcohol!: The Science and Stories of Auto-Brewery Syndrome*. Bookbaby.

Cordell, B. J., A. Kanodia. 2015. Auto-Brewery as an Emerging Syndrome: Three Representative Case Studies. *J Clin Med Case Reports* 2 (2): 5–9.

Cordell, B. J., A. Kanodia, and G. K. Miller. 2019. Case-Control Research Study of Auto-Brewery Syndrome. *Glob Adv Health Med* 8: 2164956119837566.

Cordes, E. H. 2014. *Hallelujah Moments: Tales of Drug Discovery*. New York, NY: Oxford University Press.

Crocq, M. A. Historical and cultural aspects of man's relationship with addictive drugs. 2007. *Dialogues Clin Neurosci* 9 (4): 355–61.

Currie, C. R. 2001. A community of ants, fungi, and bacteria: a multilateral approach to studying symbiosis. *Annu Rev Microbiol* 55: 357–80.

da Fonseca Pacheco, D., A. C. Freitas, A. M. Pimenta, I. D. Duarte, and M. E. de Lima. 2016. A spider derived peptide, PnPP-19, induces central antinociception mediated by opioid and cannabinoid systems. *J Venom Anim Toxins Incl Trop Dis* 22: 34.

Dallmann, R., A. Okyar, F. Lévi. 2016. Dosing-Time Makes the Poison: Circadian Regulation and Pharmacotherapy. *Trends Mol Med* 22 (5): 430–445.

Darwin, C., E. O. Wilson (editor). 2005. *From So Simple a Beginning: Darwin's Four Great Books* (Voyage of the Beagle, The Origin of Species, The Descent of Man, The Expression of Emotions in Man and Animals). New York, NY: W. W. Norton & Company.

Dashko, S., N. Zhou, C. Compagno, J. Piškur 2014. Why, when, and how did yeast evolve alcoholic fermentation? *FEMS Yeast Res* 14 (6): 826–832.

David, J., H. Merçot, P. Capy, S. McEvey, and J. Van Herrewege. 1986. Alcohol tolerance and Adh gene frequencies in European and African populations of *Drosophila melanogaster*. *Genet Sel Evol* 18 (4): 405–16.

Dawkins, R. 1995. *River Out of Eden: A Darwinian View of Life.* Science Masters Series. New York: Basic Books.

Dawkins, R. 2016. *The Selfish Gene: 40th Anniversary Edition.* Oxford University Press.

Dawkins, R., J. R. Krebs. 1979. Arm Races Between and Within Species. *Proceedings of the Royal Society of London B* 205: 489–511.

Dawson, A. G., P. Heidari, S. R. Gadagkar, M. J. Murray, and G. B. Call. 2013. An airtight approach to the inebriometer: from construction to application with volatile anesthetics. *Fly (Austin)* 7 (2): 112–7.

de las Casas, B. 1875. *Historia de las Indias.* Madrid: Impr. de M. Ginesta.

De Nobrega, A. K., A. P. Mellers, L. C. Lyons. 2017. Aging and circadian dysfunction increase alcohol sensitivity and exacerbate mortality in *Drosophila melanogaster. Exp Gerontol* 97: 49–59. doi: 10.1016/j.exger.2017.07.014.

De Nobrega, A. K., L. C. Lyons. 2016. Circadian Modulation of Alcohol-Induced Sedation and Recovery in Male and Female *Drosophila. J Biol Rhythms* 31 (2): 142–60.

De Nobrega, A. K., L. C. Lyons. 2017. *Drosophila*: An Emergent Model for Delineating Interactions between the Circadian Clock and Drugs of Abuse. *Neural Plast* 2017: 4723836.

De Nobrega, A. K., L. C. Lyons. 2018. Aging and the clock: Perspective from flies to humans. *Eur Neurosci* 2018 Sep 30.

de Pasquale, A. 1984. Pharmacognosy: the oldest modern science. *J Ethnopharmacol* 11 (1): 1–16.

de Roode, J. C., T. Lefèvre, M. D. Hunter. 2013. Ecology. Self-medication in animals. *Science* 340 (6129): 150–1.

De Simone, S. B. S., L. Q. L. Hirano, A. L. Q. Santos. 2017. Effects of midazolam in different doses in redtail boa *Boa constrictor* linnaeus, 1758 (*Squamata: Boidae*). *Cienc Anim Bras* 18 (1): e22230.

de Waal, F. 2017. *Are We Smart Enough to Know How Smart Animals Are?* New York, NY: W. W. Norton.

Degenhardt, J. et al. 2003. Attracting friends to feast on foes: engineering terpene emission to make crop plants more attractive to herbivore enemies. *Curr Opin Biotechnol* 14 (2): 169–76.

Del Toro, I., R. R. Ribbons, S. L. Pelini. 2012. The little things that run the world revisited: a review of ant-mediated ecosystem services and disservices (*Hymenoptera: Formicidae*) *Myrmecological News* 17: 133–146.

Deslauriers, J., M. Toth, A. Der-Avakian, and V. B. Risbrough. 2018. Current Status of Animal Models of Posttraumatic Stress Disorder: Behavioral and Biological Phenotypes, and Future Challenges in Improving Translation. *Biol Psychiatry* 83 (10): 895–907.

Dethier, V. G., 1962. *To Know a Fly.* San Francisco, CA: Holden-Day.

Devineni, A. V., and U. Heberlein. 2012. Acute ethanol responses in *Drosophila* are sexually dimorphic. *Proc Natl Acad Sci USA* 109 (51): 21087–92.

Dicke, M. 2009. Behavioural and community ecology of plants that cry for help. *Plant Cell Environ* 32 (6): 654–65.

Dokoumetzidis, A., and P. Macheras. 2009. Fractional kinetics in drug absorption and disposition processes. *J Pharmacokinet Pharmacodyn* 36 (2): 165–78.

Domino, E. F., E. Hornbach, T. Demana. 1993. The nicotine content of common vegetables. *N Engl J Med* 329 (6): 437.

Dominoni, D. M., S. Åkesson, R. Klaassen, K. Spoelstra, M. Bulla. 2017. Methods in field chronobiology. *Philos Trans R Soc Lond B Biol Sci* 372 (1734). pii: 20160247.

Dominy, N. J. 2015. Ferment in the family tree. *Proc Natl Acad Sci USA* 112 (2): 308–9.

Dorus, S., E. J. Vallender, P. D. Evans, J. R. Anderson, S. L. Gilbert, M. Mahowald, G. J. Wyckoff, C. M. Malcom, B. T. Lahn. 2004. Accelerated evolution of nervous system genes in the origin of Homo sapiens. *Cell* 119 (7): 1027–40.

Dossey, A. T. 2010. Insects and their chemical weaponry: new potential for drug discovery. *Nat Prod Rep* 27 (12): 1737–57.

Drouin, G., J. R. Godin, and B. Pagé. 2011. The genetics of vitamin C loss in vertebrates. *Curr Genomics* 12 (5): 371–8.

du Rand, E. E. et al. 2017. The metabolic fate of nectar nicotine in worker honey bees. *J Insect Physiol* 98: 14–22.

Dudley, R. 2000. Evolutionary origins of human alcoholism in primate frugivory. *Quarterly Review of Biology* 75: 3–15.

Dudley, R. 2002. Fermenting fruit and the historical ecology of ethanol ingestion: is alcoholism in modern humans an evolutionary hangover? *Addiction* 97: 381–388.

Dudley, R. 2014. *The Drunken Monkey: Why We Drink and Abuse Alcohol.* Berkeley, CA: University of California Press.

Dudley, S. A. 2015. Plant cooperation. *AoB Plants* 7. pii: plv113.

Duff, J. P., J. P. Holmes, P. Streete. 2012. Suspected ethanol toxicity in juvenile blackbirds and redwings. *Vet Rec* 171 (18): 453.

Duke, S. O., J. Bajsa, Z. Pan. 2013. Omics methods for probing the mode of action of natural and synthetic phytotoxins. *J Chem Ecol* 39 (2): 333–47.

Dunlap, J. C., J. J. Loros. 2016. Yes, circadian rhythms actually do affect almost everything. *Cell Res* 26 (7): 759–60.

Dunlap, J. C., J. J. Loros. 2017. Making Time: Conservation of Biological Clocks from Fungi to Animals. *Microbiol Spectr* 5 (3): 10.1128/microbiolspec.FUNK-0039-2016.

Edenberg, H. J. 2007. The genetics of alcohol metabolism: role of alcohol dehydrogenase and aldehyde dehydrogenase variants. *Alcohol Res Health* 30 (1): 5–13.

Edenberg, H. J. and J. N. McClintick. 2018. Alcohol Dehydrogenases, Aldehyde Dehydrogenases, and Alcohol Use Disorders: A Critical Review. *Alcohol Clin Exp Res* 42 (12): 2281–2297.

Edsinger, E. and G. Dölen. 2018. A Conserved Role for Serotonergic Neurotransmission in Mediating Social Behavior in Octopus. *Curr Biol* 28 (19): 3136–3142.e4.

Ehrlich, P. R., P. H. Raven. 1964. Butterflies and Plants: A Study in Coevolution. *Evolution* 18 (4): 586–608.

Ehrlich, P. R., P. H. Raven. 1967. Butterflies and plants. *Scientific American* 216: 104–113.

Eisner, T. 1964. Catnip: Its Raison D' Etre. *Science* 146 (3649): 1318–20.

Elkhawas, Y. A., A. M. Elissawy, M. S. Elnaggar, N. M. Mostafa, E. M. Kamal, M. M. Bishr, A. N. B. Singab, O. M. Salama. 2020. Chemical Diversity in Species Belonging to Soft Coral Genus *Sacrophyton* and Its Impact on Biological Activity: A Review. *Mar Drugs* 18 (1): 41.

Emerich, B. L., R. C. Ferreira, M. N. Cordeiro, M. H. Borges, A. M. Pimenta, S. G. Figueiredo, I. D. Duarte, and M. E. de Lima. 2016. δ-Ctenitoxin-Pn1a, a Peptide from *Phoneutria nigriventer* Spider Venom, Shows Antinociceptive Effect Involving Opioid and Cannabinoid Systems, in Rats. *Toxins (Basel)* 8 (4): 106.

Engel, C. 2002. Wildlife Health Care. *Canadian Wildlife* 8(3): 10–15.

Engel, C. 2003. *Wild Health.* New York, NY: Houghton Mifflin.

Engel, C. E. 2007. Zoopharmacognosy. In *Veterinary Herbal Medicine,* Wynn, S. G. and Fougere, B. J. (Eds.), 7–15. St. Louis, MO: Mosby (Elsevier) Pub.

Engleman, E. A., S. N. Katner, and B. S. Neal-Beliveau. 2016. *Caenorhabditis elegans* as a Model to Study the Molecular and Genetic Mechanisms of Drug Addiction. *Prog Mol Biol Transl Sci* 137: 229–52.

Engleman, E. A., K. B. Steagall, K. E. Bredhold, M. Breach, H. L. Kline, R. L. Bell, S. N. Katner, and B. S. Neal-Beliveau. 2018. *Caenorhabditis elegans* Show Preference for Stimulants and Potential as a Model Organism for Medications Screening. *Front Physiol* 9: 1200.

Entler, B. V., J. T. Cannon, and M. A. Seid. 2016. Morphine addiction in ants: a new model for self-administration and neurochemical analysis. *J Exp Biol* 219 (Pt 18): 2865–2869.

Erickson, C. K. 2018. *The Science of Addiction: From Neurobiology to Treatment.* New York, NY: W. W. Norton.

Eriksson, K., H. Nummi. 1983. Alcohol accumulation from ingested berries and alcohol metabolism in passerine birds. *Ornis Fenn* 60: 2–9.

Espín-Iturbe, L. T. et al. 2003. Active and passive responses to catnip (*Nepeta cataria*) are affected by age, sex and early gonadectomy in male and female cats. *Phytochemistry* 62 (5): 651–6.

Evans, J. D., M. Spivak. 2009. Socialized medicine: Individual and communal disease barriers in honey bees. *Journal of Invertebrate Pathology* 103 (1): 310–321.

Evans, W. C. 2009. *Trease and Evans' Pharmacognosy* 16th edition. Saunders.

Fagernes, C. E., K. O. Stensløkken, Å. K. Røhr, M. Berenbrink, S. Ellefsen, and G. E. Nilsson. 2017. Extreme anoxia tolerance in crucian carp and goldfish through neofunctionalization of duplicated genes creating a new ethanol-producing pyruvate decarboxylase pathway. *Sci Rep* 7 (1): 7884.

Fan, P., D. S. Manoli, O. M. Ahmed, Y. Chen, N. Agarwal, S. Kwong, A. G. Cai, J. Neitz, A. Renslo, B. S. Baker, and N. M. Shah. 2013. Genetic and neural mechanisms that inhibit *Drosophila* from mating with other species. *Cell* 154 (1): 89–102.

Farooq, M. et al. 2011. The role of allelopathy in agricultural pest management. *Pest Manag Sci* 67 (5): 493–506.

Feduccia, A. A., J. Holland, and M. C. Mithoefer. 2018. Progress and promise for the MDMA drug development program. *Psychopharmacology (Berl)* 235 (2): 561–571.

Ferchmin, P. A., O. R. Pagán, H. Ulrich, A. C. Szeto, R. M. Hann, V. A. Eterovic. 2009. Actions of octocoral and tobacco cembranoids on nicotinic receptors. *Toxicon* 54 (8): 1174–82.

Ferrea, S., G. Winterer. 2009. Neuroprotective and neurotoxic effects of nicotine. *Pharmacopsychiatry* 42 (6): 255–65.

Fishman, R. 2011. Paolo Mantegazza and Local Anesthesia. *American Journal of Ophthalmology* Volume 161, 213.

Firn, R. 2010. *Nature's Chemicals: The Natural Products That Shaped Our World.* Oxford: Oxford University Press.

Fitzgerald, S. D., J. M. Sullivan, R. J. Everson. 1990. Suspected ethanol toxicosis in two wild cedar waxwings. *Avian Dis* 34 (2): 488–90.

Flanagin, V. L. et al. 2017. Human Exploration of Enclosed Spaces through Echolocation. *J Neurosci* 37 (6): 1614–1627.

Flandreau, E. I., and M. Toth. 2018. Animal Models of PTSD: A Critical Review. *Curr Top Behav Neurosci* 38: 47–68.

Flicker, N. R., K. Poveda, H. Grab. 2019. The Bee Community of *Cannabis sativa* and Corresponding Effects of Landscape Composition. *Environ Entomol* 2019 Dec 2. pii: nvz141.

Florea, S., D. G. Panaccione, C. L. Schardl. 2017. Ergot Alkaloids of the Family *Clavicipitaceae. Phytopathology* 107 (5): 504–518.

Floru, L., J. Ishay, S. Gitter. 1969. The Influence of Psychotropic Substances on Hornet Behaviour in Colonies of *Vespa orientalis F (Hymenoptera). Psychopharmacologia* 14: 323–41.

Ford, J., C. I. Abramson, N. Sears, and F. Gutierrez. 2004. A low-cost drinkometer circuit suitable for insects and other organisms. *Psychol Rep* 94 (3 Pt 2): 1137–43.

Fossler, M. L. 1936. The Death of Hundreds of Cedar-Waxwings. *Science* 83 (2147): 185–186.

Foster, R., L. Kreitzman. 2017. *Circadian Rhythms: A Very Short Introduction* (Very Short Introductions). Oxford University Press.

Fox, M. W. 1971. Psychopathology in man and lower animals. *J Am Vet Med Assoc* 159 (1): 66–77.

Frazer, K. M., Q. Richards, D. R. Keith. 2018. The long-term effects of cocaine use on cognitive functioning: A systematic critical review. *Behav Brain Res* 348: 241–262.

Freitas, A. C., D. F. Pacheco, M. F. Machado, A. K. Carmona, I. D. Duarte, and M. E. de Lima. 2016. PnPP-19, a spider toxin peptide, induces peripheral antinociception through opioid and cannabinoid receptors and inhibition of neutral endopeptidase. *Br J Pharmacol* 173 (9): 1491–501.

Frischknecht, H. R. and P. G. Waser. 1978a. Actions of hallucinogens on ants (*Formica pratensis*)-I. Brain levels of LSD and the following oral administration. *Gen Pharmacol* 9 (5): 369–73.

Frischknecht, H. R. and P. G. Waser. 1978b. Actions of hallucinogens on ants (*Formica pratensis*)-II. Effects of amphetamine, LSD and delta-9-tetrahydrocannabinol. *Gen Pharmacol* 9 (5): 375–80.

Frischknecht, H. R. and P. G. Waser. 1980. Actions of hallucinogens on ants (*Formica pratensis*)-III. Social behavior under the influence of LSD and tetrahydrocannabinol. *Gen Pharmacol* 11 (1): 97–106.

Fry, J. D. 2014. Mechanisms of naturally evolved ethanol resistance in *Drosophila melanogaster. J Exp Biol* 217 (Pt 22): 3996–4003.

Fry, J. D., K. Donlon, and M. Saweikis. 2008. A worldwide polymorphism in aldehyde dehydrogenase in *Drosophila melanogaster*: evidence for selection mediated by dietary ethanol. *Evolution* 62 (1): 66–75.

Fry, J. D., and M. Saweikis. 2006. Aldehyde dehydrogenase is essential for both adult and larval ethanol resistance in *Drosophila melanogaster. Genet Res* 87 (2): 87–92.

Fu, J., H. Liu, H. Xing, H. Sun, Z. Ma, B. Wu. 2014. Comparative analysis of glucuronidation of ethanol in treeshrews, rats and humans. *Xenobiotica* 44 (12): 1067–73.

Fu, X., P. Giavalisco, X. Liu, G. Catchpole, N. Fu, Z. B. Ning, S. Guo, Z. Yan, M. Somel, S. Pääbo S, et al. 2011. Rapid metabolic evolution in human prefrontal cortex. *Proc Natl Acad Sci USA* 108 (15): 6181–6.

Gagliano, M. 2013. Seeing Green: The Re-discovery of Plants and Nature's Wisdom. *Societies* 3: 147–157.

Gagliano, M. 2014. In a green frame of mind: perspectives on the behavioural ecology and cognitive nature of plants. *AoB Plants* 7.

Gagliano, M. 2018. *Thus Spoke the Plant: A Remarkable Journey of Groundbreaking Scientific Discoveries and Personal Encounters with Plants.* Berkeley, CA: North Atlantic Books.

Gagliano, M. et al. 2014. Experience teaches plants to learn faster and forget slower in environments where it matters. *Oecologia* 175 (1): 63–72.

Gagliano, M. et al. 2016. Learning by Association in Plants. *Sci Rep* 6: 38427.

Gagliano, M. et al. 2017. Tuned in: plant roots use sound to locate water. *Oecologia* 184 (1): 151–160.

Gagliano, M., C. I. Abramson, M. Depczynski. 2018. Plants learn and remember: let's get used to it. *Oecologia* 186 (1): 29–31.

Ganswindt, A., S. Muenscher, M. Henley, S. Henley, M. Heistermann, R. Palme, P. Thompson, H. Bertschinger. 2010. Endocrine correlates of musth and the impact of ecological and social factors in free-ranging African elephants (*Loxodonta africana*). *Horm Behav* 57 (4–5): 506–14.

García-Diez, M., D. Garrido, D. Hoffmann, P. Pettitt, A. Pike, J. Zilhão. 2015. The chronology of hand stencils in European Palaeolithic rock art: implications of new U-series results from El Castillo Cave (Cantabria, Spain). *J Anthropol Sci* 93: 135–52.

García Pardo, M. P. et al. 2017. Animal models of drug addiction. *Adicciones* 29 (4): 278–292.

Garnatje, T., J. Peñuelas, J. Vallès. 2017. Reaffirming 'Ethnobotanical Convergence.' *Trends Plant Sci* 22 (8): 640–641.

Gochman, S. R., M. B. Brown, and N. J. Dominy. 2016. Alcohol discrimination and preferences in two species of nectar-feeding primate. *R Soc Open Sci* 3 (7): 160217.

Godfrey-Smith, P. 2016. *Other Minds: The Octopus, the Sea, and the Deep Origins of Consciousness.* New York, NY: Farrar, Straus and Giroux.

Goode, J. 2014. Intoxicating science. *Nature* 509: 286.

Gregg, J. 2013. *Are Dolphins Really Smart?: The mammal behind the myth.* Oxford: Oxford University Press.

Grémiaux, A. et al. 2014. Plant anesthesia supports similarities between animals and plants: Claude Bernard's forgotten studies. *Plant Signal Behav* 9 (1): e27886.

Greven, D. 2009. *Gender and Sexuality in Star Trek: Allegories of Desire in the Television Series and Films.* Jefferson, NC: McFarland & Company.

Grienke, U. et al. 2014. European medicinal polypores-a modern view on traditional uses. *J Ethnopharmacol* 154 (3): 564–83.

Griffin, D. R. 1981. *The Question of Animal Awareness.* New York, NY: Rockefeller University Press.

Griffin, D. R. 2001. *Animal Minds: Beyond Cognition to Consciousness.* Chicago, IL: Chicago University Press.

Grisel, J. 2020. *Never Enough: The Neuroscience and Experience of Addiction.* NY: Penguin Random House.

Groot, A. T., T. Dekker, and D. G. Heckel. 2016. The Genetic Basis of Pheromone Evolution in Moths. *Annu Rev Entomol* 61: 99–117.

Grover, N. 1965. Man and Plants against Pain. *Economic Botany* 19 (2): 99–112.

Grüter, C. and W. M. Farina. 2009. The honeybee waggle dance: can we follow the steps? *Trends Ecol Evol* 24 (5): 242–7.

Gutnick, T. and M. J. Kuba. 2018. Animal Behavior: Socializing Octopus. *Curr Biol* 28 (19): R1147–R1149.

Guevara-Fiore, P. and J. A. Endler. 2014. Male sexual behaviour and ethanol consumption from an evolutionary perspective: A comment on Sexual Deprivation Increases Ethanol Intake in *Drosophila. Fly (Austin)* 8 (4): 234–6.

Guo, X., W. Zhang, J. Ma, Z. Liu, D. Hu, J. Chen et al. 2018. The case study of one patient with gut fermentation syndrome: case report and review of the literature. *Int J Clin Exp Med* 11 (4): 4324–9.

Haarmann, T., Y. Rolke, S. Giesbert, P. Tudzynski. 2009. Ergot: from witchcraft to biotechnology. *Mol Plant Pathol* 10 (4): 563–77.

Hafez, E. M., M. A. Hamad, M. Fouad, and A. Abdel-Lateff. 2017. Auto-brewery syndrome: Ethanol pseudo-toxicity in diabetic and hepatic patients. *Hum Exp Toxicol* 36 (5): 445–450.

Hagen, E. H. et al. 2009. Ecology and neurobiology of toxin avoidance and the paradox of drug reward. *Neuroscience* 160 (1): 69–84.

Hagen, E. H., C. J. Roulette, R. J. Sullivan. 2013. Explaining human recreational use of 'pesticides': The neurotoxin regulation model of substance use vs. the hijack model and implications for age and sex differences in drug consumption. *Front Psychiatry* 4: 142.

Hanifin, C. T. 2010. The chemical and evolutionary ecology of tetrodotoxin (TTX) toxicity in terrestrial vertebrates. *Mar Drugs* 8 (3): 577–93.

Hardin, P. E. 2000. From biological clock to biological rhythms. *Genome Biol* 1 (4): reviews1023.

Hardy, K. et al. 2012. Neanderthal medics? Evidence for food, cooking, and medicinal plants entrapped in dental calculus. *Naturwissenschaften* 99 (8): 617–26.

Hardy, K., S. Buckley, L. Copeland. 2018. Pleistocene dental calculus: Recovering information on Paleolithic food items, medicines, paleoenvironment and microbes. *Evol Anthropol* 27 (5): 234–246.

Harper, J. L. 1975. Allelopathy. *Quarterly Review in Biology* 50: 493–495.

Harris, A. 2019. *Conscious: A Brief Guide to the Fundamental Mystery of the Mind*. New York, NY: HarperCollins.

Hayman, J. 2019. Charles Darwin. *RCS Bull* 101 (3): 100–104.

Hedges, S. B. et al. 2004. A molecular timescale of eukaryote evolution and the rise of complex multicellular life. *BMC Evol Biol* 4: 2.

Hedges, S. B. et al. 2015. Tree of life reveals clock-like speciation and diversification. *Mol Biol Evol* 32 (4): 835–45.

Hedges, S. B. et al. 2018. Accurate timetrees require accurate calibrations. *Proc Natl Acad Sci USA* 115 (41): E9510–E9511.

Heinrich, M. 2003. Ethnobotany and natural products: the search for new molecules, new treatments of old diseases or a better understanding of indigenous cultures? *Curr Top Med Chem* 3 (2): 141–54.

Heinrich, M., S. Anagnostou. 2017. From Pharmacognosia to DNA-Based Medicinal Plant Authentication— Pharmacognosy through the Centuries. *Planta Med* 83 (14–15): 1110–1116.

Helfand, W. H. 1980. Vin Mariani. *Pharmacy in History* 22 (1): 11–19.

Helfrich-Förster, C. 2005. Neurobiology of the fruit fly's circadian clock. *Genes Brain Behav* 4 (2): 65–76. Review.

Helfrich-Förster, C. 2019. Light input pathways to the circadian clock of insects with an emphasis on the fruit fly *Drosophila melanogaster*. *J Comp Physiol A*. doi:10.1007/s00359-019-01379-5.

Henry, J. M., D. Cring. 2013. Geophagy: An Anthropological Perspective. In *Soils and Human Health*, edited by Eric C. Brevik and Lynn C. Burgess, 179–198. CRC Press Taylor & Francis Group.

Hernández-Tobías, A., A. Julián-Sánchez, E. Piña, and H. Riveros-Rosas. 2011. Natural alcohol exposure: is ethanol the main substrate for alcohol dehydrogenases in animals? *Chem Biol Interact* 191 (1–3): 14–25.

Hesselberg, T., and F. Vollrath. 2004. The effects of neurotoxins on web-geometry and web-building behaviour in *Araneus diadematus* Cl. *Physiol Behav* 82 (2–3): 519–29.

Heyer, D. B., and R. M. Meredith. 2017. Environmental toxicology: Sensitive periods of development and neurodevelopmental disorders. *Neurotoxicology* 58: 23–41.

Hjelmqvist, L., A. Norin, M. El-Ahmad, W. Griffiths, and H. Jörnvall. 2003. Distinct but parallel evolutionary patterns between alcohol and aldehyde dehydrogenases: addition of fish/human betaine aldehyde dehydrogenase divergence. *Cell Mol Life Sci* 60 (9): 2009–16.

Hoang, K. et al. 2017. Host Diet Affects the Morphology of Monarch Butterfly Parasites. *Journal of Parasitology* 103 (3): 228–236.

Hochner, B. 2012. An embodied view of octopus neurobiology. *Curr Biol* 22 (20): R887–92.

Hochner, B. 2013. How nervous systems evolve in relation to their embodiment: what we can learn from octopuses and other molluscs. *Brain Behav Evol* 82 (1): 19–30.

Hochner, B., D. L. Glanzman. 2016. Evolution of highly diverse forms of behavior in molluscs. *Curr Biol* 26 (20): R965–R971.

Hofmann, A. 2009. *LSD, My Problem Child: Reflections on Sacred Drugs, Mysticism and Science*. MAPS.org.

Hoffmann, D. L., D. E. Angelucci, V. Villaverde, J. Zapata, J. Zilhão. 2018. Symbolic use of marine shells and mineral pigments by Iberian Neandertals 115,000 years ago. *Sci Adv* 4 (2): eaar5255.

Hoffmann, D. L., C. D. Standish, A. W. G. Pike, M. García-Diez, P. B. Pettitt, D. E. Angelucci, V. Villaverde, J. Zapata, J. A. Milton, J. Alcolea-González J., et al. 2018. Dates for Neanderthal art and symbolic behaviour are reliable. *Nat Ecol Evol* 2 (7): 1044–1045.

Hoffmann, D. L., C. D. Standish, M. García-Diez, P. B. Pettitt, J. A. Milton, J. Zilhão, J. J. Alcolea-González, P. Cantalejo-Duarte, H. Collado, R. de Balbín, et al. 2018. U-Th dating of carbonate crusts reveals Neandertal origin of Iberian cave art. *Science* 359 (6378): 912–915.

Hokama, Y. et al. 1990. Causitive toxin(s) in the death of two Atlantic dolphins. *J Clin Lab Anal* 4 (6): 474–8.

Hölldobler, B., and E. O. Wilson. 2008. *The Superorganism: The Beauty, Elegance, and Strangeness of Insect Societies.* New York, NY: W. W. Norton & Company.

Hölldobler, B., and E. O. Wilson. 2010. *The Leafcutter Ants: Civilization by Instinct.* New York: W. W. Norton & Company.

Hopper, S. D., and H. Lambers. 2009. Darwin as a plant scientist: a Southern Hemisphere perspective. *Trends Plant Sci* 14 (8): 421–35.

Horgan, C., G. Terrero, and G. Wessel. 2011. In the extremes-traumatic insemination. *Mol Reprod Dev* 78 (5): Fmi.

Houston, D. M., L. L. Head. 1993. Acute alcohol intoxication in a dog. *Can Vet J* 34 (1): 41–42.

Huber, R., A. Imeh-Nathaniel, T. I. Nathaniel, S. Gore, U. Datta, R. Bhimani, J. B. Panksepp, J. Panksepp, and M. J. van Staaden. 2018. Drug-sensitive Reward in Crayfish: Exploring the Neural Basis of Addiction with Automated Learning Paradigms. *Behav Processes* 152: 47–53.

Huber, R., J. B. Panksepp, T. Nathaniel, A. Alcaro, and J. Panksepp. 2011. Drug-sensitive reward in crayfish: an invertebrate model system for the study of seeking, reward, addiction, and withdrawal. *Neurosci Biobehav Rev* 35 (9): 1847–53.

Huffman, M. A. 1997. Current Evidence for Self-Medication in Primates: A Multidisciplinary Perspective. *Yearbook of Physical Anthropology* 40: 171–200.

Huffman, M. A. 2001. Self-Medicative Behavior in the African Great Apes: An Evolutionary Perspective into the Origins of Human Traditional Medicine. *BioScience* 51 (8): 651–661.

Huffman, M. A. 2003. Animal self-medication and ethno-medicine: exploration and exploitation of the medicinal properties of plants. *Proc Nutr Soc* 62 (2): 371–81.

Hughes, W. O., A. N. Bot, and J. J. Boomsma. 2010. Caste-specific expression of genetic variation in the size of antibiotic-producing glands of leaf-cutting ants. *Proc Biol Sci* 277 (1681): 609–15.

Hwang, D. F., T. Noguchi. 2007. Tetrodotoxin poisoning. *Adv Food Nutr Res* 52: 141–236.

Hyland Bruno, J., O. Tchernichovski. 2019. Regularities in zebra finch song beyond the repeated motif. *Behav Processes* 163: 53–59.

Ibrahim, U. H., N. Devnarain, T. Govender. 2021. Biomimetic strategies for enhancing synthesis and delivery of antibacterial nanosystems. *Int J Pharm* 596: 120276.

Inderjit, S.O. Duke. 2003. Ecophysiological aspects of allelopathy. *Planta* 217 (4): 529–39.

Inderjit et al. 2011. The ecosystem and evolutionary contexts of allelopathy. *Trends Ecol Evol* 26 (12): 655–62.

Iriti, M. 2013. Plant neurobiology, a fascinating perspective in the field of research on plant secondary metabolites. *Int J Mol Sci* 14 (6): 10819–21.

Islam, M. 2011. *Tracing the Evolutionary History of Coca (Erythroxylum).* PhD Dissertation, University of Colorado at Boulder.

Jackson, D. E. 2009. Nutritional ecology: a first vegetarian spider. *Curr Biol* 19 (19): R894–5.

Jaffe, J. H. 1990. Drug addiction and drug abuse. In *Goodman and Gilman's the Pharmacological Basis of Therapeutics*, Goodman, A. G., T. W. Rall, A. S. Nies, and P. Taylor eds, 522–573. New York: McGraw Hill.

Jain, C. P. et al. 2008. Animal self-medication through natural sources. *Natural Product Rad* 7: 49–53.

Janecka, J. E., W. Miller, T. H. Pringle, F. Wiens, A. Zitzmann, K. M. Helgen, M. S. Springer, W. J. Murphy. 2007. Molecular and genomic data identify the closest living relative of primates. *Science* 318 (5851): 792–4.

Janzen, D. H. 1978. Complications in interpreting the chemical defenses of trees against tropical arboreal plant-eating vertebrates. In *The Ecology of Arboreal Folivores*, G.G. Montgomery, ed., 73–84. Washington, DC: Smithsonian Institution Press.

Jentzsch, H. C. 1983. Of elephants and psychiatry. *Freedom* 58: 6–7.

Jeong, K. J. et al. 2012. Effective control of slug damage through tobacco extract and caffeine solution in combination with alcohol. *Horticulture, Environment, and Biotechnology* 53 (2): 123–128.

Ji, H. F., X. J. Li, H. Y. Zhang. 2009. Natural products and drug discovery. Can thousands of years of ancient medical knowledge lead us to new and powerful drug combinations in the fight against cancer and dementia? *EMBO Rep* 10 (3): 194–200.

Jimenez-Del-Rio, M., A. Daza-Restrepo, and C. Velez-Pardo. 2008. The cannabinoid CP55,940 prolongs survival and improves locomotor activity in *Drosophila melanogaster* against paraquat: implications in Parkinson's disease. *Neurosci Res* 61 (4): 404–11.

Jones, P. L. et al. 2019. Cardenolide Intake, Sequestration, and Excretion by the Monarch Butterfly along Gradients of Plant Toxicity and Larval Ontogeny. *J Chem Ecol* 45 (3): 264–277.

Kacsoh, B. Z., Z. R. Lynch, N. T. Mortimer, and T. A. Schlenke. 2013. Fruit flies medicate offspring after seeing parasites. *Science* 339 (6122): 947–50.

Kaji, H., Y. Asanuma, H. Ide, N. Saito, M. Hisamura, M. Murao, T. Yoshida, K. Takahashi. 1976. The auto-brewery syndrome—the repeated attacks of alcoholic intoxication due to the overgrowth of *Candida (albicans)* in the gastrointestinal tract. *Mater Med Pol* 8 (4): 429–35.

Kalan, A. K. et al. 2019. Chimpanzees use tree species with a resonant timbre for accumulative stone throwing. *Biol Lett* 15 (12): 20190747.

Kalske, A. et al. 2019. Insect Herbivory Selects for Volatile-Mediated Plant-Plant Communication. *Curr Biol* 29 (18): 3128-3133.

Kandel, E. R. 1989. Genes, nerve cells, and the remembrance of things past. *J Neuropsychiatry Clin Neurosci* 1 (2): 103–25.

Karczmar, A. G. 1993. Brief presentation of the story and present status of studies of the vertebrate cholinergic system. *Neuropsychopharmacology* 9 (3): 181–99.

Kaul, P. N., P. Daftari. 1986. Marine pharmacology: bioactive molecules from the sea. *Annu Rev Pharmacol Toxicol* 26: 117–42.

Kaun, K. R., A. V. Devineni, and U. Heberlein. 2012. *Drosophila melanogaster* as a model to study drug addiction. *Hum Genet* 131 (6): 959–75.

Keller, M. 1979. A historical overview of alcohol and alcoholism. *Cancer Res* 39 (7 Pt 2): 2822–9.

Kellermann, K. I., E. N. Bouton, S. S. Brandt. 2020. Is Anyone Out There? In *Open Skies. Historical & Cultural Astronomy*. Springer.

Kelly, D. R. 1996. When is a butterfly like an elephant? *Chemistry & Biology* 3: 595–602.

Kennedy, D. O. 2014. *Plants and the Human Brain*. Oxford, UK: Oxford University Press.

Kessler, A., I. T. Baldwin. 2002. Plant responses to insect herbivory: the emerging molecular analysis. *Annu Rev Plant Biol* 53: 299–328.

Kinde, H., E. Foate, E. Beeler, F. Uzal, J. Moore, R. Poppenga. 2012. Strong circumstantial evidence for ethanol toxicosis in Cedar Waxwings (*Bombycilla cedrorum*). *J Ornithol* 153: 995–998.

King, S. L., S. J. Allen, M. Krützen, R. C. Connor. 2019. Vocal behaviour of allied male dolphins during cooperative mate guarding. *Anim Cogn* 22 (6): 991–1000.

Knuepfer, M. M. 2003. Cardiovascular disorders associated with cocaine use: myths and truths. *Pharmacology & Therapeutics* 97: 181–222.

Kruckenberg, K. M., A. F. DiMartini, J. A. Rymer, A. W. Pasculle, K. Tamama. 2020. Urinary Auto-brewery Syndrome: A Case Report. *Ann Intern Med*. doi: 10.7326/L19-0661.

Kuczaj, S. A., and H. C. Skelinen. 2014. Why do Dolphins Play? *Animal Behavior and Cognition* 1 (2): 113–127.

Kuczaj, S. A., and D. B. Yeater. 2007. Observations of rough-toothed dolphins (Steno bredanensis) off the coast of Utila, Honduras. *J Mar Biol Ass* UK 87: 141–148.

Kuhar, M. 2015. *The Addicted Brain: Why We Abuse Drugs, Alcohol, and Nicotine*. Pearson.

Kühl, H. S. et al. 2016. Chimpanzee accumulative stone throwing. *Sci Rep* 6: 22219.

Kuhlman, S. J., L. M. Craig, J. F. Duffy. 2018. Introduction to Chronobiology. *Cold Spring Harb Perspect Biol* 10 (9). pii: a033613.

Kumar, S. et al. 2017. TimeTree: A Resource for Timelines, Timetrees, and Divergence Times. *Mol Biol Evol* 34 (7): 1812–1819.

Kutschera, U., and W. R. Briggs. 2009. From Charles Darwin's botanical country-house studies to modern plant biology. *Plant Biol (Stuttg)* 11 (6): 785–95.

Kutschera, U., and K. J. Niklas. 2009. Evolutionary plant physiology: Charles Darwin's forgotten synthesis. *Naturwissenschaften* 96 (11): 1339–54.

La Plante, M. D. 2019. *Superlative: The Biology of Extremes*. Houston, TX: BenBella Books.

Laaksonen, J., T. Laaksonen, J. Itämies, S. Rytkönen, P. Välimäki. 2006. A new efficient bait-trap model for Lepidoptera surveys—the Oulu model. *Entomol Fennica* 17: 153–160.

Lago, J., L. P. Rodríguez, L. Blanco, J. M. Vieites, A. G. Cabado. 2015. Tetrodotoxin, an Extremely Potent Marine Neurotoxin: Distribution, Toxicity, Origin and Therapeutical Uses. *Mar Drugs* 13 (10): 6384–406.

Larkins, N., S. Wynn. 2004. Pharmacognosy: phytomedicines and their mechanisms. *Vet Clin North Am Small Anim Pract* 34 (1): 291–327.

Larsson, S. A., L. Backlund, L. Bohlin. 2008. Reappraising a decade old explanatory model for pharmacognosy. *Phytochemistry Letters* 1: 131–134.

Leão, P. N., M. T. Vasconcelos, and V. M. Vasconcelos. 2009. Allelopathy in freshwater cyanobacteria. *Crit Rev Microbiol* 35 (4): 271–82.

Lee, A. H., C. L. Brandon, J. Wang, W. N. Frost. 2018. An Argument for Amphetamine Induced Hallucinations in an Invertebrate. *Front Physiol* 9: 730.

Lee, H. G., Y. C. Kim, J. S. Dunning, and K. A. Han. 2008. Recurring ethanol exposure induces disinhibited courtship in *Drosophila*. *PLoS One* 3 (1): e1391.

Leroi-Gourhan, A. 1975. The Flowers Found with Shanidar IV, a Neanderthal Burial in Iraq. *Science* 190 (4214): 562–564.

Levey, D. J., M. L. Cipollini. 1998. A Glycoalkaloid in Ripe Fruit Deters Consumption by Cedar Waxwings. *The Auk* 115 (2): 359–367.

Levin, M., A. M. Pietak, and J. Bischof. 2019. Planarian regeneration as a model of anatomical homeostasis: Recent progress in biophysical and computational approaches. *Semin Cell Dev Biol* 87: 125–144.

Lewin, L. 1998. *Phantastica: A Classic Survey on the Use and Abuse of Mind-Altering Plants.* Park Street Press.

Lewis, M. 2015. *The Biology of Desire: Why Addiction Is Not a Disease.* New York, NY: Perseus.

Li, Q., X. Ni. 2016. An early Oligocene fossil demonstrates treeshrews are slowly evolving living fossils. *Sci Rep* 6: 18627.

Lieberman, D. Z., Long, M. E. 2018. *The Molecule of More: How a Single Chemical in Your Brain Drives Love, Sex, and Creativity—and Will Determine the Fate of the Human Race.* Dallas, TX: BenBella Books.

Lietava, J. 1992. Medicinal plants in a Middle Paleolithic grave Shanidar IV? *J Ethnopharmacol* 35 (3): 263–6.

Lilly, J. C. 1965. Dolphin-Human Relation and LSD 25. In *The Use of LSD in Psychotherapy and Alcoholism*, Abramson, H. A., ed. New York: The Bobbs-Merrill Company.

Lilly, J. C., A. M. Miller. 1961. Vocal Exchanges between Dolphins: Bottlenose dolphins "talk" to each other with whistles, clicks, and a variety of other noises. *Science* 134 (3493): 1873–6.

Lindsay, W. L. 1879. *Mind in the lower animals in health and disease.* London UK: C. K. Paul.

Lindstedt, S. L., and K. C. Nishikawa. 2015. From Tusko to Titin: the role for comparative physiology in an era of molecular discovery. *Am J Physiol Regul Integr Comp Physiol* 308 (12): R983–9.

Little, A. E., T. Murakami, U. G. Mueller, and C. R. Currie. 2006. Defending against parasites: fungus-growing ants combine specialized behaviours and microbial symbionts to protect their fungus gardens. *Biol Lett* 2 (1): 12–6.

Liu, T., L. Dartevelle, C. Yuan, H. Wei, Y. Wang, J. F. Ferveur, and A. Guo. 2009. Reduction of dopamine level enhances the attractiveness of male *Drosophila* to other males. *PLoS One* 4 (2): e4574.

Logan, B. K., and A. W. Jones. 2000. Endogenous ethanol 'auto-brewery syndrome' as a drunk-driving defence challenge. *Med Sci Law* 40 (3): 206–15.

Lorenz, N., T. Haarmann, S. Pazoutová, M. Jung, P. Tudzynski. 2009. The ergot alkaloid gene cluster: functional analyses and evolutionary aspects. *Phytochemistry* 70 (15–16): 1822–32.

Lowenstein, E. G., and N. A. Velazquez-Ulloa. 2018. A Fly's Eye View of Natural and Drug Reward. *Front Physiol* 9: 407.

Maderspacher, F. 2008. Otzi. *Curr Biol* 18 (21): R990–1.

Maher, C. 2019. Experiment Rather Than Define. *Trends Plant Sci* 25 (3): 213–14.

Malik, F., P. Wickremesinghe, J. Saverimuttu. 2019. Case report and literature review of auto-brewery syndrome: probably an underdiagnosed medical condition. *BMJ Open Gastroenterol* 6 (1): e000325.

Malve, H. 2016. Exploring the ocean for new drug developments: Marine pharmacology. *J Pharm Bioallied Sci* 8 (2): 83–91.

Mancuso, S., A. Viola. 2018. *Brilliant Green: The Surprising History and Science of Plant Intelligence.* Washington, DC: Island Press.

Margotta, V., B. Caronti, G. Meco, A. Merante, S. Ruggieri, G. Venturini, and G. Palladini. 1997. Effects of cocaine treatment on the nervous system of planaria (*Dugesia gonocephala* s. l.). Histochemical and ultrastructural observations. *Eur J Histochem* 41 (3): 223–30.

Marino, L. Convergence of complex cognitive abilities in cetaceans and primates. 2002. *Brain Behav Evol* 59 (1–2): 21–32.

Markel, H. 2011. Über coca: Sigmund Freud, Carl Koller, and cocaine. *JAMA* 305 (13): 1360–1.

Marquet, P. A., R. A. Quiñones, S. Abades, F. Labra, M. Tognelli, M. Arim, M. Rivadeneira. 2005. Scaling and power-laws in ecological systems. *J Exp Biol* 208 (Pt 9): 1749–69.

Mather, J. A. 2008. Cephalopod consciousness: behavioural evidence. *Conscious Cogn* 17 (1): 37–48.

Mather, J. A. 2011. Philosophical background of attitudes toward and treatment of invertebrates. *ILAR J* 52 (2): 205–12.

Mather, J. A, L. Dickel. 2017. Cephalopod complex cognition. *Current Opinion in Behavioral Sciences* 16: 131–137.

Mayer, A. M., K. R. Gustafson. 2008. Marine pharmacology in 2005-2006: antitumour and cytotoxic compounds. *Eur J Cancer* 44 (16): 2357–87.

Mayer, A. M. S., A. D. Rodríguez, O. Taglialatela-Scafati, N. Fusetani. 2017. Marine Pharmacology in 2012-2013: Marine Compounds with Antibacterial, Antidiabetic, Antifungal, Anti-Inflammatory, Antiprotozoal, Antituberculosis, and Antiviral Activities; Affecting the Immune and Nervous Systems, and Other Miscellaneous Mechanisms of Action. *Mar Drugs* 15 (9). pii: E273.

McArt, S. H. et al. 2014. Arranging the bouquet of disease: floral traits and the transmission of plant and animal pathogens. *Ecol Lett* 17 (5): 624–36.

McConnell, J. V. 1965. *A manual of psychological experimentation on planarians.* Self-published.

McGovern, P. E. 2009. Homo Imbibens: I Drink, Therefore I Am. In *Uncorking the Past: The Quest for Wine, Beer, and Other Alcoholic Beverages.* Los Angeles, CA: University of California Press.

McKenna, D. J. 1996. Plant hallucinogens: springboards for psychotherapeutic drug discovery. *Behav Brain Res* 73 (1–2): 109–16. Review.

McKenna, T. 1993. *Food of the Gods: The Search for the Original Tree of Knowledge, a Radical History of Plants, Drugs, and Human Evolution.* Bantam Books.

McPartland, J., V. Di Marzo, L. De Petrocellis, A. Mercer, and M. Glass. 2001. Cannabinoid receptors are absent in insects. *J Comp Neurol* 436 (4): 423–9.

McPartland, J. M. 2004. Phylogenomic and chemotaxonomic analysis of the endocannabinoid system. *Brain Res Brain Res Rev* 45 (1): 18–29.

McPartland, J. M. 2018. Systematics at the Levels of Family, Genus, and Species. *Cannabis Cannabinoid Res* 3 (1): 203–212.

McPartland, J. M., J. Agraval, D. Gleeson, K. Heasman, and M. Glass. 2006. Cannabinoid receptors in invertebrates. *J Evol Biol* 19 (2): 366–73.

McPartland, J. M., and M. Glass. 2003. Functional mapping of cannabinoid receptor homologs in mammals, other vertebrates, and invertebrates. *Gene* 312: 297–303.

McPartland, J. M., I. Matias, V. Di Marzo, and M. Glass. 2006. Evolutionary origins of the endocannabinoid system. *Gene* 370: 64–74.

Mendl, M. et al. 2011. Animal behaviour: emotion in invertebrates? *Curr Biol* 21 (12): R463–5.

Merçot, H. 1994. Phenotypic expression of ADH regulatory genes in *Drosophila melanogaster*: a comparative study between a paleartic and a tropical population. *Genetica* 94 (1): 37–41.

Merçot, H., D. Defaye, P. Capy, E. Pla, and J. R. David. 1994. Alcohol tolerance, adh activity, and ecological niche of *Drosophila* species. *Evolution* 48 (3): 746–757.

Merlin, M. D. 2003. Archaeological Evidence for the Tradition of Psychoactive Plant Use in the Old World. *Economic Botany* 57 (3): 295–323.

Michel, M., L. C. Lyons. 2014. Unraveling the complexities of circadian and sleep interactions with memory formation through invertebrate research. *Front Syst Neurosci* 8: 133.

Michels, J., S. N. Gorb, and K. Reinhardt. 2015. Reduction of female copulatory damage by resilin represents evidence for tolerance in sexual conflict. *J R Soc Interface* 12 (104): 20141107.

Milan, N. F., B. Z. Kacsoh, and T. A. Schlenke. 2012. Alcohol consumption as self-medication against blood-borne parasites in the fruit fly. *Curr Biol* 22 (6): 488–93.

Milton, K. 2004. Ferment in the family tree: does a frugivorous dietary heritage influence contemporary patterns of human ethanol use? *Integr Comp Biol* 44 (4): 304–14.

Misof, B. et al. 2014. Phylogenomics resolves the timing and pattern of insect evolution. *Science* 346 (6210): 763–7.

Mohammed, A. N., N. Alugubelly, B. L. Kaplan, and R. L. Carr. 2018. Effect of repeated juvenile exposure to Δ9-tetrahydrocannabinol on anxiety-related behavior and social interactions in adolescent rats. *Neurotoxicol Teratol* 69: 11–20.

Mohr, S. E. 2018. *First in Fly: Drosophila Research and Biological Discovery.* Cambridge, MA: Harvard University Press.

Molyneux, R. J., and L. F. James. 1982. Loco intoxication: Indolizidine alkaloids of spotted locoweed (*Astragalus lentiginosus*). *Science* 216: 190–191.

Montgomery, S. 2016. *The Soul of an Octopus: A Surprising Exploration into the Wonder of Consciousness.* Atria Books.

Morris, S., D. Humphreys, and D. Reynolds. 2006. Myth, marula, and elephant: an assessment of voluntary ethanol intoxication of the African elephant (*Loxodonta africana*) following feeding on the fruit of the marula tree (*Sclerocarya birrea*). *Physiol Biochem Zool* 79 (2): 363–9.

Müller, C. P. 2018. Animal models of psychoactive drug use and addiction—Present problems and future needs for translational approaches. *Behav Brain Res* 352: 109–115.

Musgrave, S. et al. 2019. Teaching varies with task complexity in wild chimpanzees. *Proc Natl Acad Sci USA.* pii: 201907476.

Musgrave, S. et al. 2016. Tool transfers are a form of teaching among chimpanzees. *Sci Rep* 6: 34783.

Mustard, J. A. 2014. The buzz on caffeine in invertebrates: effects on behavior and molecular mechanisms. *Cell Mol Life Sci* 71 (8): 1375–82.

Mustard, J. A., L. Dews, A. Brugato, K. Dey, and G. A. Wright. 2012. Consumption of an acute dose of caffeine reduces acquisition but not memory in the honey bee. *Behav Brain Res* 232 (1): 217–24.

Nagel, T. 1974. What Is It Like to Be a Bat? *The Philosophical Review* 83 (4): 435–450.

Nagel, T. 2012. *Mind & Cosmos: Why the Materialist Neo-Darwinian Conception of Nature is Almost Certainly False.* New York, NY: Oxford University Press.

Narahashi, T. 2008. Tetrodotoxin: a brief history. *Proc Jpn Acad Ser B Phys Biol Sci* 84 (5): 147–54.

Nässel, D. R., and C. Wegener. 2011. A comparative review of short and long neuropeptide F signaling in invertebrates: Any similarities to vertebrate neuropeptide Y signaling? *Peptides* 32 (6): 1335–55.

Nathanson, J. A. et al. 1993. Cocaine as a naturally occurring insecticide. *Proc Natl Acad Sci USA* 90 (20): 9645–8.

Nayak, S., A. Roberts, K. Bires, C. S. Tallarida, E. Kim, M. Wu, and S. M. Rawls. 2016. Benzodiazepine inhibits anxiogenic-like response in cocaine or ethanol withdrawn planarians. *Behav Pharmacol* 27 (6): 556–8.

Nesee, R. M. 2019. *Good Reasons for Bad Feelings: Insights from the Frontier of Evolutionary Psychiatry.* New York, NY: Dutton.

Nesse, R. M., K. C. Berridge. 1997. Psychoactive drug use in evolutionary perspective. *Science* 278 (5335): 63–6.

Nesse, R. M., and G. C. Williams. 1998. Evolution and the origins of disease. *Sci Am* 279 (5): 86–93.

Nestorović, J., D. Misić, B. Siler, M. Soković, J. Glamočlija, A. Cirić, V. Maksimović, D. Grubišić. 2010. Nepetalactone content in shoot cultures of three endemic *Nepeta* species and the evaluation of their antimicrobial activity. *Fitoterapia* 81 (6): 621–6.

Nevill, A. M., A. D. Stewart, T. Olds, R. Holder. 2004. Are adult physiques geometrically similar? The dangers of allometric scaling using body mass power laws. *Am J Phys Anthropol* 124 (2): 177–82.

Ng, J. Y., S. Obuobi, M. L. Chua, C. Zhang, S. Hong, Y. Kumar, R. Gokhale, P. L. R. Ee. 2020. Biomimicry of microbial polysaccharide hydrogels for tissue engineering and regenerative medicine—A review. *Carbohydr Polym* 241: 116345.

Ni, X., G. Gunawan, S. L. Brown, P. E. Sumner, J. R. Ruberson, G. D. Buntin, C. C. Holbrook, R. D. Lee, D. A. Streett, J. E. Throne, and J. F. Campbell. 2008. Insect-attracting and antimicrobial properties of antifreeze for monitoring insect pests and natural enemies in stored corn. *J Econ Entomol* 101 (2): 631–6.

Nichols, D. E. 2014. The Heffter Research Institute: past and hopeful future. *J Psychoactive Drugs* 46 (1): 20–6.

Nichols, D. E. 2016. Psychedelics. *Pharmacol Rev* 68 (2): 264–355.

Nichols, G. E. 1919. The general biology course and the teaching of elementary botany and zoology in American colleges and universities. *Science* 50: 509–517.

Nürnberger, F., A. Keller, S. Härtel, and I. Steffan-Dewenter. 2019. Honey bee waggle dance communication increases diversity of pollen diets in intensively managed agricultural landscapes. *Mol Ecol* 28 (15): 3602–3611.

Nürnberger, F., I. Steffan-Dewenter, and S. Härtel. 2017. Combined effects of waggle dance communication and landscape heterogeneity on nectar and pollen uptake in honey bee colonies. *PeerJ* 5: e3441.

Nzabanita, C., H. Liu, S. Min, M. Ting-yan, and Y. Z. Li. 2018. Locoweed Endophytes: A Review. *J Plant Physiol Pathol* 6: 5.

Oberhaus, D. 2019. *Extraterrestrial Languages*. Cambridge: The MIT Press.

O'Grady, P. M., and R. DeSalle. 2018. Phylogeny of the Genus *Drosophila*. *Genetics* 209 (1): 1–25.

Ogueta, M., O. Cibik, R. Eltrop, A. Schneider, and H. Scholz. 2010. The influence of Adh function on ethanol preference and tolerance in adult *Drosophila melanogaster*. *Chem Senses* 35 (9): 813–22.

Ohdo, S., S. Koyanagi, N. Matsunaga. 2019. Chronopharmacological strategies focused on chrono-drug discovery. *Pharmacol Ther* 202: 72–90.

Ojelade, S. A., and A. Rothenfluh. 2009. Addiction: flies hit the skids. *Curr Biol* 19 (24): R1110–1.

Olson, C.R., D. C. Owen, A. E. Ryabinin, C. V. Mello. 2014. Drinking songs: alcohol effects on learned song of zebra finches. *PLoS One* 9 (12): e115427.

Orsolini, L. et al. 2018. Psychedelic Fauna for Psychonaut Hunters: A Mini-Review. *Front Psychiatry* 9: 153.

Orbach, D. N., N. Veselka, Y. Dzal, L. Lazure, and M. B. Fenton. 2010. Drinking and flying: does alcohol consumption affect the flight and echolocation performance of phyllostomid bats? *PLoS One* 5 (2): e8993.

Osmond, H. 1957. A review of the clinical effects of psychotomimetic agents. *Ann NY Acad Sci* 66: 418–434.

Pagán, O. R. 1998. Effects of cembranoids and anesthetic agents on [$^3$H]-tenocyclidine binding to the nicotinic acetylcholine receptor from *Torpedo californica*. San Juan, PR: University of Puerto Rico, MS Thesis.

Pagán, O. R. 2005. *Synthetic local anesthetics as alleviators of cocaine inhibition of the human dopamine transporter*. Ithaca, NY: Cornell University, PhD Dissertation.

Pagán, O. R. 2014. *The First Brain: The Neuroscience of Planarians*. New York: Oxford University Press.

Pagán, O. R. 2017. Planaria: an animal model that integrates development, regeneration and pharmacology. *Int J Dev Biol* 61 (8–9): 519–529.

Pagán, O. R. 2018. *Strange Survivors: How Organisms Attack and Defend in the Game of Life*. Dallas, TX: BenBella Books.

Pagán, O. R. 2019. The brain: a concept in flux. *Philos Trans R Soc Lond B Biol Sci* 374 (1774): 20180383.

Pagán, O. R., D. Baker, S. Deats, E. Montgomery, M. Tenaglia, C. Randolph, D. Kotturu, C. Tallarida, D. Bach, G. Wilk, S. Rawls, and R. B. Raffa. 2012. Planarians in pharmacology: parthenolide is a specific behavioral antagonist of cocaine in the planarian *Girardia tigrina*. *Int J Dev Biol* 56 (1–3): 193–6.

Pagán, O. R., T. Coudron, and T. Kaneria. 2009. The flatworm planaria as a toxicology and behavioral pharmacology animal model in undergraduate research experiences. *J Undergrad Neurosci Educ* 7 (2): A48–52.

Pagán, O. R., S. Deats, D. Baker, E. Montgomery, G. Wilk, M. Tenaglia, and J. Semon. 2013. Planarians require an intact brain to behaviorally react to cocaine, but not to react to nicotine. *Neuroscience* 246: 265–70.

Pagán, O. R., E. Montgomery, S. Deats, D. Bach, and D. Baker. 2015. Evidence of Nicotine-Induced, Curare-Insensitive, Behavior in Planarians. *Neurochem Res* 40 (10): 2087–90.

Pagán, O. R., A. L. Rowlands, M. Azam, K. R. Urban, A. H. Bidja, D. M. Roy, R. B. Feeney, and L. K. Afshari. 2008. Reversal of cocaine-induced planarian behavior by parthenolide and related sesquiterpene lactones. *Pharmacol Biochem Behav* 89 (2): 160–70.

Pagán, O. R., A. L. Rowlands, A. L. Fattore, T. Coudron, K. R. Urban, A. H. Bidja, and V. A. Eterović. 2009. A cembranoid from tobacco prevents the expression of nicotine-induced withdrawal behavior in planarian worms. *Eur J Pharmacol* 615 (1–3): 118–24.

Pagán, O. R., A. L. Rowlands, and K. R. Urban. 2006. Toxicity and behavioral effects of dimethylsulfoxide in planaria. *Neurosci Lett* 407 (3): 274–8.

Palladini, G., S. Ruggeri, F. Stocchi, M. F. De Pandis, G. Venturini, and V. Margotta. 1996. A pharmacological study of cocaine activity in planaria. *Comp Biochem Physiol C Pharmacol Toxicol Endocrinol* 115 (1): 41–5.

Pallen, M. 2009. *The Rough Guide to Evolution*. London, UK: Rough Guides.

Palmer, J. D. 2002. *The Living Clock: The Orchestrator of Biological Rhythms*. Oxford University Press.

Pandey, H. P., A. K. Verma. 2017. A study on the role of holy basil (*Ocimum sanctum*) in auto-healing of Indian garden lizard (*Calotes versicolor*). *International Journal of Fauna and Biological Studies* 4 (2): 97–100.

Papagianni, D., M. A. Morse. 2015. *The Neanderthals Rediscovered: How Modern Science is Rewriting Their Story*. New York, NY: Thames and Hudson.

Park, A., A. Ghezzi, T. P. Wijesekera, and N. S. Atkinson. 2017. Genetics and genomics of alcohol responses in *Drosophila*. *Neuropharmacology* 122: 22–35.

Pascual-Garrido, A. 2019. Cultural variation between neighbouring communities of chimpanzees at Gombe, Tanzania. *Sci Rep* 9 (1): 8260.

Patterson, I. A., R. J. Reid, B. Wilson, K. Grellier, H. M. Ross, P. M. Thompson. 1998. Evidence for infanticide in bottlenose dolphins: an explanation for violent interactions with harbour porpoises? *Proc Biol Sci* 265 (1402): 1167–70.

Pendergrast, M. 2013. *For God, Country, and Coca-Cola: The Definitive History of the Great American Soft Drink and the Company That Makes It*. Basic Books.

Peñaflor, M. F., M. Bento. 2013. Herbivore-induced plant volatiles to enhance biological control in agriculture. *Neotrop Entomol* 42 (4): 331–43.

Perry, C. J., and A. B. Barron. 2013. Neural mechanisms of reward in insects. *Annu Rev Entomol* 58: 543–62.

Pesce, S., O. Perceval, C. Bonnineau, C. Casado-Martinez, A. Dabrin, E. Lyautey, E. Naffrechoux, B. J. D. Ferrari. 2018. Looking at biological community level to improve ecotoxicological assessment of freshwater sediments: report on a first French-Swiss workshop. *Environ Sci Pollut Res Int* 25 (1): 970–974.

Petschenka, G. 2016. How herbivores coopt plant defenses: natural selection, specialization, and sequestration. *Curr Opin Insect Sci* 14: 17–24.

Petschenka, G. et al. 2017. Convergently Evolved Toxic Secondary Metabolites in Plants Drive the Parallel Molecular Evolution of Insect Resistance. *Am Nat* 190 (S1): S29–S43.

Petschenka, G. et al. 2018. Relative Selectivity of Plant Cardenolides for Na. *Front Plant Sci* 9: 1424.

Petschenka, G., A. A. Agrawal. 2015. Milkweed butterfly resistance to plant toxins is linked to sequestration, not coping with a toxic diet. *Proc Biol Sci* 282 (1818): 20151865.

Peterson, K. J., J. A. Cotton, J. G. Gehling, D. Pisani. 2008. The Ediacaran emergence of bilaterians: congruence between the genetic and the geological fossil records. *Philos Trans R Soc Lond B Biol Sci* 363 (1496): 1435–43.

Pettersson, L. B., M. Franzén. 2008. Comparing wine-based and beer-based baits for moth trapping: a field experiment. *Entomologisk Tidskrift* 129 (3): 129–134.

Petruccelli, E., and K. R. Kaun. 2019. Insights from intoxicated *Drosophila*. *Alcohol* 74: 21–27.

Pfiester, M., P. G. Koehler, R. M. Pereira. 2009. Sexual conflict to the extreme: traumatic insemination in bed bugs. *Am Entomol* 55: 244–249.

Píchová, K., S. Pažoutová, M. Kostovčík, M. Chudíčková, E. Stodůlková, P. Novák, M. Flieger, E. van der Linde, M. Kolařík. 2018. Evolutionary history of ergot with a new infrageneric classification (*Hypocreales: Clavicipitaceae: Claviceps*). *Mol Phylogenet Evol* 123: 73–87.

Piechulla, B., M. C. Lemfack, M. Kai. 2017. Effects of discrete bioactive microbial volatiles on plants and fungi. *Plant Cell Environ* 40 (10): 2042–2067.

Pigeault, R. et al. 2016. Evolution of transgenerational immunity in invertebrates. *Proc Biol Sci* 283 (1839).

Pike, A.W., D. L. Hoffman, M. García-Diez, P. B. Pettitt, J. Alcolea, R. De Balbín, C. González-Sainz, C. de las Heras, J. A. Lasheras, R. Montes, J. Zilhão. 2012. *Science* 336 (6087): 1409–13.

Poinar, G., S. Alderman, J. Wunderlich. 2015. One-hundred-million-year-old ergot: psychotropic compounds in the Cretaceous? *Palaeodiversity* 8: 13–19.

Ponnappa, B. C., E. Rubin. 2000. Modeling alcohol's effects on organs in animal models. *Alcohol Res Health* 24 (2): 93–104.

Porter, J. W., N. M. Targett. 1988. Allelochemical Interactions between Sponges and Corals. *Biological Bulletin* 175 (2): 230–239.

Prasad, V. et al. 2005. Dinosaur coprolites and the early evolution of grasses and grazers. *Science* 310 (5751): 1177–80.

Pratheepa, V., V. Vasconcelos. 2017. Binding and Pharmacokinetics of the Sodium Channel Blocking Toxins (Saxitoxin and the Tetrodotoxins). *Mini Rev Med Chem* 17 (4): 320–327.

Queller, D. C., J. E. Strassmann. 2003. Eusociality. *Current Biology* 13 (22): R861–3.

Quik, M., K. O'Leary, C. M. Tanner. 2008. Nicotine and Parkinson's disease: implications for therapy. *Mov Disord* 23 (12): 1641–52.

Raffa, R. B., P. Desai. 2005. Description and quantification of cocaine withdrawal signs in Planaria. *Brain Res* 1032 (1–2): 200–2.

Raffa, R. B., S. Rawls. 2008. *Planaria: a model for drug action and abuse* (Molecular Biology Intelligence Unit). Austin, TX: Landes Bioscience.

Rains, G. C., J. K. Tomberlin, and D. Kulasiri. 2008. Using insect sniffing devices for detection. *Trends Biotechnol* 26 (6): 288–94.

Rains, G., D. Kulasiri, Z. Zhou, S. Samarasinghe, J. Tomberlin, and D. Olson. 2010. Synthesizing neurophysiology, genetics, behaviour and learning to produce whole-insect programmable sensors to detect volatile chemicals. *Biotechnol Genet Eng Rev* 26: 179–204.

Raman, R., S. Kandula. 2008. Zoopharmacognosy. *Resonance* 13 (3): 245–253.

Rasmann, S., A. A. Agrawal. 2011. Latitudinal patterns in plant defense: evolution of cardenolides, their toxicity and induction following herbivory. *Ecol Lett* 14 (5): 476–83.

Rasmussen, S. 2014. *The Quest for Aqua Vitae: The History and Chemistry of Alcohol from Antiquity to the Middle Ages* (SpringerBriefs in Molecular Science), 60–63. New York, NY: Springer.

Raubenheimer, D., S. J. Simpson. 2009. Nutritional PharmEcology: Doses, nutrients, toxins, and medicines. *Integr Comp Biol* 49 (3): 329–37.

Rawls, S. M., K. Gerber, Z. Ding, C. Roth, and R. B. Raffa. 2008. Agmatine: identification and inhibition of methamphetamine, kappa opioid, and cannabinoid withdrawal in planarians. *Synapse* 62 (12): 927–34.

Reddien, P. W. 2018. The Cellular and Molecular Basis for Planarian Regeneration. *Cell* 175 (2): 327–345.

Reed, C. F., P. N. Witt, and M. B. Scarboro. 1982. Maturation and d-amphetamine-induced changes in web building. *Dev Psychobiol* 15 (1): 61–70.

Rehm, H., D. Gradmann. 2010. Plant neurobiology, intelligent plants or stupid studies. *Lab Times* March: 30–32.

Reppert, S. M., J. C. de Roode. 2018. Demystifying Monarch Butterfly Migration. *Curr Biol* 28 (17): R1009–R1022.

Reppert, S. M., R. J. Gegear, and C. Merlin. 2010. Navigational mechanisms of migrating monarch butterflies. *Trends Neurosci* 33 (9): 399–406.

Reppert, S. M., P. A. Guerra, and C. Merlin. 2016. Neurobiology of Monarch Butterfly Migration. *Annu Rev Entomol* 61: 25–42.

Reynolds, H. T. et al. 2018. Horizontal gene cluster transfer increased hallucinogenic mushroom diversity. *Evol Lett* 2 (2): 88–101.

Richardson, L. L. et al. 2015. Secondary metabolites in floral nectar reduce parasite infections in bumblebees. *Proc Biol Sci* 282 (1803): 20142471.

Rightmire, G. P. 2004. Brain size and encephalization in early to Mid-Pleistocene Homo. *Am J Phys Anthropol* 124 (2): 109–23.

Ritson-Williams, R., M. Yotsu-Yamashita, V. J. Paul. 2006. Ecological functions of tetrodotoxin in a deadly polyclad flatworm. *Proc Natl Acad Sci USA* 103 (9): 3176–9.

Rodríguez, A. D. 1995. The natural products chemistry of West Indian gorgonian octocorals. *Tetrahedron* 51 (16): 4571–4618.

Rodríguez, E., R. Wrangham. 1993. Zoopharmacognosy: The Use of Medicinal Plants by Animals. In *Phytochemical Potential of Tropical Plants. Recent Advances in Phytochemistry* (Proceedings of the Phytochemical Society of North America), Downum, K. R., J. T. Romeo, H. A. Stafford (eds), vol 27, 89–105. Boston, MA: Springer.

Rogers, A. R., N. S. Harris, A. A. Achenbach. 2020. Neanderthal-Denisovan ancestors interbred with a distantly related hominin. *Sci Adv* 6 (8): eaay5483.

Roth, O. et al. 2018. Recent advances in vertebrate and invertebrate transgenerational immunity in the light of ecology and evolution. *Heredity (Edinb)* 121 (3): 225–238.

Rowlands, A. L., and O. R. Pagán. 2008. Parthenolide prevents the expression of cocaine-induced withdrawal behavior in planarians. *Eur J Pharmacol* 583 (1): 170–2.

Ryan, F. 2011. *The Mystery of Metamorphosis: A Scientific Detective Story*. Chelsea Green Publishing.

Ryvkin, J., A. Bentzur, S. Zer-Krispil, and G. Shohat-Ophir. 2018. Mechanisms Underlying the Risk to Develop Drug Addiction, Insights from Studies in *Drosophila melanogaster*. *Front Physiol* 9: 327.

Sachs, J. L., U. G. Mueller, T. P. Wilcox, J. J. Bull. 2004. The evolution of cooperation. *Q Rev Biol* 79 (2): 135–60. Review.

Safina, C. 2016. *Beyond Words*. New York, NY: Picador.

Sagan, C. 2000. *The Cosmic Connection: An Extraterrestrial Perspective 2nd Edition*. Cambridge: Cambridge University Press.

Samorini, G. 1995. Paolo Mantegazza (1831-1910): Italian pioneer in the studies on drugs. *Eleusis* 2: 14–20.

Samorini, G. 2002. *Animals and Psychedelics: The Natural World and the Instinct to Alter Consciousness*. Rochester, VT: Park Street Press.

Samu, F., F. Vollrath. 1992. Spider orb web as bioassay for pesticide side effects. *Entomol Exp Appl* 62: 117–124.

Sánchez, F., C. Korine, B. P. Kotler, and B. Pinshow. 2008. Ethanol concentration in food and body condition affect foraging behavior in Egyptian fruit bats (*Rousettus aegyptiacus*). *Naturwissenschaften* 95 (6): 561–7.

Sánchez, F., B. P. Kotler, C. Korine, and B. Pinshow. 2008. Sugars are complementary resources to ethanol in foods consumed by Egyptian fruit bats. *J Exp Biol* 211 (Pt 9): 1475–81.

Sánchez, F., M. Melcón, C. Korine, and B. Pinshow. 2010. Ethanol ingestion affects flight performance and echolocation in Egyptian fruit bats. *Behav Processes* 84 (2): 555–8.

Saniotis, A. 2010. Evolutionary and anthropological approaches towards understanding human need for psychotropic and mood altering substances. *J Psychoactive Drugs* 42 (4): 477–84.

Sapolsky, R. 2004. *Why Zebras Don't Get Ulcers*. New York, NY: Holt.

Sattelle, D. B., and S. D. Buckingham. 2006. Invertebrate studies and their ongoing contributions to neuroscience. *Invert Neurosci* 6 (1): 1–3.

Scheel, D., P. Godfrey-Smith, and M. Lawrence. 2016. Signal Use by Octopuses in Agonistic Interactions. *Curr Biol* 26 (3): 377–82.

Schenk, S., and D. Newcombe. 2018. Methylenedioxymethamphetamine (MDMA) in Psychiatry: Pros, Cons, and Suggestions. *J Clin Psychopharmacol* 38 (6): 632–638.

Schiff, P. L. 2002. Opium and Its Alkaloids. *American Journal of Pharmaceutical Education* Vol. 66.

Schmidt, J. O. 2016. *The Sting of the Wild*. Baltimore, MD: Johns Hopkins University Press.

Schmidt, J. O. 2018. Clinical consequences of toxic envenomations by *Hymenoptera*. *Toxicon* 150: 96–104.

Schmidt, J. O. 2019. Pain and Lethality Induced by Insect Stings: An Exploratory and Correlational Study. *Toxins (Basel)* 11 (7).

Schneider, A., M. Ruppert, O. Hendrich, T. Giang, M. Ogueta, S. Hampel, M. Vollbach, A. Büschges, and H. Scholz. 2012. Neuronal basis of innate olfactory attraction to ethanol in *Drosophila*. *PLoS One* 7 (12): e52007.

Schöner, J., A. Heinz, M. Endres, K. Gertz, and G. Kronenberg. 2017. Post-traumatic stress disorder and beyond: an overview of rodent stress models. *J Cell Mol Med* 21 (10): 2248–2256.

Schott, M., B. Klein, A. Vilcinskas. 2015. Detection of Illicit Drugs by Trained Honeybees (*Apis mellifera*). *PLoS One* 10 (6): e0128528.

Schwarz, D., D. Bloom, R. Castro, O. R. Pagán, and C. A. Jiménez-Rivera. 2011. Parthenolide Blocks Cocaine's Effect on Spontaneous Firing Activity of Dopaminergic Neurons in the Ventral Tegmental Area. *Curr Neuropharmacol* 9 (1): 17–20.

Sekhon, M. L., O. Lamina, K. E. Hogan, and C. L. Kliethermes. 2016. Common genes regulate food and ethanol intake in *Drosophila*. *Alcohol* 53: 27–34.

Sessa, B., L. Higbed, and D. Nutt. 2019. A Review of 3,4-methylenedioxymethamphetamine (MDMA)-Assisted Psychotherapy. *Front Psychiatry* 10: 138.

Shen, F., Y. Duan, S. Jin, N. Sui. 2014. Varied behavioral responses induced by morphine in the tree shrew: a possible model for human opiate addiction. *Front Behav Neurosci* 8: 333.

Shipley, G. P., K. Kindscher. 2016. Evidence for the Paleoethnobotany of the Neanderthal: A Review of the Literature. *Scientifica (Cairo)* 2016: 8927654.

Shohat-Ophir, G., K. R. Kaun, R. Azanchi, H. Mohammed, and U. Heberlein. 2012. Sexual deprivation increases ethanol intake in *Drosophila*. *Science* 335 (6074): 1351–5.

Shultz, S., E. Nelson, R. I. Dunbar. 2012. Hominin cognitive evolution: identifying patterns and processes in the fossil and archaeological record. *Philos Trans R Soc Lond B Biol Sci* 367 (1599): 2130–40.

Shurkin, J. 2014. News feature: Animals that self-medicate. *Proc Natl Acad Sci USA* 111 (49): 17339–41.

Si, A., S. W. Zhang, and R. Maleszka. 2005. Effects of caffeine on olfactory and visual learning in the honey bee (*Apis mellifera*). *Pharmacol Biochem Behav* 82 (4): 664–72.

Siegel, R. K. 1977. Cocaine: Recreational Use and Intoxication. In *Cocaine: 1977*, edited by R. C. Petersen and R. C. Stillman, 119–36. National Institute on Drug Abuse Research Monograph 13. DHEW Publication no. (ADM) 77–741. Washington, DC: US Government Printing Office.

Siegel, R. K. 1984. LSD-induced effects in elephants: Comparisons with musth behavior. *Bulletin of the Psychonomic Society*, 22 (1): 53–56.

Siegel, R. K. 2005. *Intoxication: The Universal Drive for Mind-Altering Substances*. Park Street Press.

Siegel, R. K., M. Brodie. 1984. Alcohol self-administration by elephants. *Bulletin of the Psychonomic Society* 22 (1): 49–52.

Siegel, S. 2005. Two views of the addiction elephant: comment in McSweeney, Murphy, and Kowal (2005). *Exp Clin Psychopharmacol* 13 (3): 190–3; discussion 194–9.

Simone, M., J. D. Evans, and M. Spivak. 2009. Resin collection and social immunity in honey bees. *Evolution* 63 (11): 3016–22.

Simone-Finstrom, M. et al. 2017. Propolis Counteracts Some Threats to Honey Bee Health. *Insects* 8 (2).

Simone-Finstrom, M. D., and M. Spivak. 2012. Increased resin collection after parasite challenge: a case of self-medication in honey bees? *PLoS One* 7 (3): e34601.

Singh, C. M., and U. Heberlein. 2000. Genetic control of acute ethanol-induced behaviors in *Drosophila*. *Alcohol Clin Exp Res* 24 (8): 1127–36.

Sladky, K. K., M. E. Kinney, S. M. Johnson. 2008. Analgesic efficacy of butorphanol and morphine in bearded dragons and corn snakes. *J Am Vet Med Assoc* 233 (2): 267–73.

Smith, M. L. 2014. Honey bee sting pain index by body location. *PeerJ* 2: e338.

Snoek, T., K. J. Verstrepen, and K. Voordeckers. 2016. How do yeast cells become tolerant to high ethanol concentrations? *Curr Genet* 62 (3): 475–80.

Solecki, R. 1957. Shanidar Cave. *Scientific American*, 197 (5): 58–65.

Solecki, R. 1975. Shanidar IV, a neanderthal flower burial in northern Iraq. *Science* 190: 880–881.

Soloway, S. B. 1976. Naturally occurring insecticides. *Environ Health Perspect* 14: 109–17.

Sommer, D. J. 1999. The Shanidar IV 'Flower Burial': a Re-evaluation of Neanderthal Burial Ritual. *Cambridge Archaeological Journal* 9 (1): 127–129.

Søvik, E., A. B. Barron. 2013. Invertebrate models in addiction research. *Brain Behav Evol* 82 (3): 153–65.

Søvik, E., P. Berthier, W. P. Klare, P. Helliwell, E. L. S. Buckle, J. A. Plath, A. B. Barron, and R. Maleszka. 2018. Cocaine Directly Impairs Memory Extinction and Alters Brain DNA Methylation Dynamics in Honey Bees. *Front Physiol* 9: 79.

Søvik, E., J. L. Cornish, and A. B. Barron. 2013. Cocaine tolerance in honey bees. *PLoS One* 8 (5): e64920.

Søvik, E., N. Even, C. W. Radford, and A. B. Barron. 2014. Cocaine affects foraging behaviour and biogenic amine modulated behavioural reflexes in honey bees. *PeerJ* 2: e662.

Spanagel, R. 2017. Animal models of addiction. *Dialogues Clin Neurosci* 19 (3): 247–258.

Sparks, J. T., J. D. Bohbot, M. Ristic, D. Mišic, M. Skoric, A. Mattoo, J. C. Dickens. 2017. Chemosensory Responses to the Repellent *Nepeta* Essential Oil and Its Major Component Nepetalactone by *Aedes aegypti* (*Diptera: Culicidae*), a Vector of Zika Virus. *J Med Entomol* 54 (4): 957–963.

Spivak, M., M. Goblirsch, M. Simone-Finstrom. 2019. Social-medication in bees: the line between individual and social regulation. *Curr Opin Insect Sci* 33: 49–55.

Staaf, D. 2020. *Monarchs of the Sea: The Extraordinary 500-Million-Year History of Cephalopods.* New York: The Experiment.

Stahlberg, R. 2006. Historical overview on plant neurobiology. *Plant Signal Behav* 1 (1): 6–8.

Staub, P. O., M. S. Geck, C. S. Weckerle, L. Casu, and M. Leonti. 2015. Classifying diseases and remedies in ethnomedicine and ethnopharmacology. *J Ethnopharmacol* 174: 514–9.

Steele, E. J., S. Al-Mufti, K. A. Augustyn, R. Chandrajith, J. P. Coghlan, S. G. Coulson, S. Ghosh, M. Gillman, R. M. Gorczynski, B. Klyce et al. 2018. Cause of Cambrian Explosion—Terrestrial or Cosmic? *Prog Biophys Mol Biol* 136: 3–23.

Steiner, L. 1995. Rough-toothed dolphin, Steno bredanensis: a new species record for the Azores, with some notes on behaviour. *Arquipélago; Life and Marine Sciences* 13A: 125–127.

Stephen, L. J., W. J. Walley. 2000. Alcohol intoxication contributing to mortality in Bohemian waxwings and a pine grosbeak. *Blue Jay* 58 (1): 33–35.

Stephens, D., R. Dudley. 2004. The drunken monkey hypothesis. *Natural History* 113 (10): 40–44.

Steppuhn, A. et al. 2004. Nicotine's defensive function in nature. *PLoS Biol* 2 (8): e217.

Sternberg, E. D., J. C. de Roode, and M. D. Hunter. 2015. Trans-generational parasite protection associated with paternal diet. *J Anim Ecol* 84 (1): 310–21.

Stockley, P., and J. Bro-Jørgensen. 2011. Female competition and its evolutionary consequences in mammals. *Biol Rev Camb Philos Soc* 86 (2): 341–66.

Stockley, P., J. Bro-Jørgensen. 2011. Female competition and its evolutionary consequences in mammals. *Biol Rev Camb Philos Soc* 86 (2): 341–66.

Stolberg, V. B. 2011. The Use of Coca: Prehistory, History, and Ethnography. *Journal of Ethnicity in Substance Abuse* 10: 126–146.

Strachecka, A., M. Krauze, K. Olszewski, G. Borsuk, J. Paleolog, M. Merska, J. Chobotow, M. Bajda, and K. Grzywnowicz. 2014. Unexpectedly strong effect of caffeine on the vitality of western honeybees (*Apis mellifera*). *Biochemistry (Mosc)* 79 (11): 1192–201.

Struempf, H. M., J. E. Schondube, C. Martínez Del Rio. 1999. The Cyanogenic Glycoside Amygdalin Does Not Deter Consumption of Ripe Fruit by Cedar Waxwings. *The Auk* 116 (3): 749–758.

Suárez-Rodríguez, M., I. López-Rull, C. M. Garcia. 2012. Incorporation of cigarette butts into nests reduces nest ectoparasite load in urban birds: new ingredients for an old recipe? *Biol Lett* 9 (1): 20120931.

Suárez-Rodríguez, M., C. Macías Garcia. 2014. There is no such a thing as a free cigarette; lining nests with discarded butts brings short-term benefits, but causes toxic damage. *J Evol Biol* 27 (12): 2719–26.

Sullivan, R. J., E. H. Hagen. 2002. Psychotropic substance-seeking: evolutionary pathology or adaptation? *Addiction* 97 (4): 389–400.

Sullivan, R. J., E. H. Hagen, P. Hammerstein. 2008. Revealing the paradox of drug reward in human evolution. *Proc Biol Sci* 275 (1640): 1231–41.

Sweatt, J. D. 2016. Neural plasticity and behavior—sixty years of conceptual advances. *J Neurochem* 139 Suppl 2: 179–199.

Tabakoff, B., P. L. Hoffman. 2000. Animal models in alcohol research. *Alcohol Res Health* 24 (2): 77–84.

Tahara, Y., S. Shibata. 2013. Chronobiology and nutrition. *Neuroscience* 253: 78–88.

Taiz, L. et al. 2019. Plants Neither Possess nor Require Consciousness. *Trends Plant Sci* 24 (8): 677–687.

Tallarida, C. S., K. Bires, J. Avershal, R. J. Tallarida, S. Seo, S. M. Rawls. 2014. Ethanol and cocaine: environmental place conditioning, stereotypy, and synergism in planarians. *Alcohol* 48 (6): 579–86.

Tamaki, N., T. Morisaka, T. Michihiro. 2006. Does body contact contribute towards repairing relationships? The association between flipper-rubbing and aggressive behavior in captive bottlenose dolphins. *Behav Processes* 73 (2): 209–15.

Tan, W. H. et al. 2018. The Effects of Milkweed Induced Defense on Parasite Resistance in Monarch Butterflies, *Danaus plexippus. J Chem Ecol* 44 (11): 1040–1044.

Tanford, C., J. Reynolds. 2004. *Nature's Robots: A History of Proteins.* New York, NY: Oxford University Press.

Tao, L. et al. 2016. Fitness costs of animal medication: antiparasitic plant chemicals reduce fitness of monarch butterfly hosts. *J Anim Ecol* 85 (5): 1246–54.

Tetreau, G. et al. 2019. Trans-generational Immune Priming in Invertebrates: Current Knowledge and Future Prospects. *Front Immunol* 10: 1938.

Thaler, L. et al. 2019. Human Click-Based Echolocation of Distance: Superfine Acuity and Dynamic Clicking Behaviour. *J Assoc Res Otolaryngol*.

Thaler, L., M. A. Goodale. 2016. Echolocation in humans: an overview. *Wiley Interdiscip Rev Cogn Sci* 7 (6): 382–393.

Thiriez, C. et al. 2011. Can nicotine be used medicinally in Parkinson's disease? *Expert Rev Clin Pharmacol* 4 (4): 429–36.

Thorburn, L. P. et al. 2015. Variable effects of nicotine, anabasine, and their interactions on parasitized bumble bees. *F1000Res* 4: 880.

Trewavas, A. et al. 2020. Consciousness Facilitates Plant Behavior. *Trends Plant Sci*. pii: S1360-1385(19)30340-1.

Tsai, P. S. 2018. Gonadotropin-releasing hormone by any other name would smell as sweet. *Gen Comp Endocrinol* 264: 58–63.

Tsatsakis, A. M. et al. 2018. The dose response principle from philosophy to modern toxicology: The impact of ancient philosophy and medicine in modern toxicology science. *Toxicol Rep* 5: 1107–1113.

Tsuchiya, H. Anesthetic Agents of Plant Origin: A Review of Phytochemicals with Anesthetic Activity. *Molecules* 22 (8). pii: E1369.

Tye, M. 2016. *Tense Bees and Shell-Shocked Crabs: Are Animals Conscious?* New York, NY: Oxford University Press.

Ursprung, H., and L. Carlin. 1968. *Drosophila* alcohol dehydrogenase: in vitro changes of isozyme patterns. *Ann N Y Acad Sci* 151 (1): 456–75.

Vallee, B. L. 1998. Alcohol in the Western World. *Scientific American* 278 (6): 80–85.

Van Dyke, C., and R. Byck. 1982. Cocaine. *Scientific American* 246 (3): 128–41.

van Renterghem, T. 1995. *When Santa Was A Shaman: Ancient Origins of Santa Claus & the Christmas Tree*. Llewellyn Publications.

Van Waarde, A., G. Van den Thillart, C. Erkelens, A. Addink, and J. Lugtenburg. 1990. Functional coupling of glycolysis and phosphocreatine utilization in anoxic fish muscle. An in vivo 31P NMR study. *J Biol Chem* 265 (2): 914–23.

Vassiliou, V. A. 2011. Effectiveness of insecticides in controlling the first and second generations of the *Lobesia botrana* (*Lepidoptera: Tortricidae*) in table grapes. *J Econ Entomol* 104 (2): 580–5.

Vemuri, V. K., and A. Makriyannis. 2015. Medicinal chemistry of cannabinoids. *Clin Pharmacol Ther* 97 (6): 553–8.

Vergara-Fernández, A. et al. 2018. Biofiltration of volatile organic compounds using fungi and its conceptual and mathematical modeling. *Biotechnol Adv* 36 (4): 1079–1093.

Vet, L.E.M., M. Dicke. 1992. Ecology of infochemical use by natural enemies in a tritrophic context. *Annu Rev Entomol* 37: 141–172.

Villella, A., and J. C. Hall. 2008. Neurogenetics of courtship and mating in *Drosophila*. *Adv Genet* 62: 67–184.

Vunduk, J., A. Klaus, M. Kozarski, P. Petrovic, Z. Zizak, M. Niksic, L. J. Van Griensven. 2015. Did the Iceman Know Better? Screening of the Medicinal Properties of the Birch Polypore Medicinal Mushroom, *Piptoporus betulinus* (Higher Basidiomycetes). *Int J Med Mushrooms* 17 (12): 1113–25.

Wahlberg, I., and A.-M. Eklund. 1992. Cembranoids, pseudopteranoids and cubitanoids of natural occurrence. *Progress in the Chemistry of Organic Natural Products* 59: 141–294.

Wall, J. D. et al. 2013. Higher levels of neanderthal ancestry in East Asians than in Europeans. *Genetics* 194 (1): 199–209.

Walters, D. 2017. *Fortress Plant: How to survive when everything wants to eat you*. New York, NY: Oxford University Press.

Wandersee, J. H. and E. E. Schussler. 1999. Preventing Plant Blindness. *The American Biology Teacher*, 61 (2): 82–86.

Wang, D. Y., S. Kumar, S. B. Hedges. 1999. Divergence time estimates for the early history of animal phyla and the origin of plants, animals and fungi. *Proc Biol Sci* 266 (1415): 163–71.

War, A. R. et al. 2011. Herbivore induced plant volatiles: their role in plant defense for pest management. *Plant Signal Behav* 6 (12): 1973–8.

Ward, P. 2018. *Lamarck's Revenge: How Epigenetics Is Revolutionizing Our Understanding of Evolution's Past and Present*. Bloomsbury Publishing.

Ward, R. R. 1971. *The Living Clocks*. Knopf.

Wasson et al. 2008. *The Road to Eleusis: Unveiling the Secret of the Mysteries*. Berkeley, CA: North Atlantic Books.

Weiner, J. 2000. *Time, Love, Memory: A Great Biologist and His Quest for the Origins of Behavior*. New York, NY. Vintage Books.

Wells, D. L., J. M. Egli. 2004. The influence of olfactory enrichment on the behaviour of captive black-footed cats, *Felis nigripes*. *Applied Animal Behaviour Science* 85: 107–119.

West, G. 2017. *Scale: The Universal Laws of Life, Growth, and Death in Organisms, Cities, and Companies*. New York, NY: Penguin Press.

West, L. J., C. M. Pierce, and W. D. Thomas. 1962. Lysergic Acid Diethylamide: Its Effects on a Male Asiatic Elephant. *Science* 138 (3545): 1100–3.

Weyrich, L. S. et al. 2017. Neanderthal behaviour, diet, and disease inferred from ancient DNA in dental calculus. *Nature* 544 (7650): 357–361.

Whiten, A. 2017. Culture extends the scope of evolutionary biology in the great apes. *Proc Natl Acad Sci USA* 114 (30): 7790–7797.

Wiens, F., A. Zitzmann, M. A. Lachance, M. Yegles, F. Pragst, F. M. Wurst, D. von Holst, S. L. Guan, R. Spanagel. 2008. Chronic intake of fermented floral nectar by wild treeshrews. *Proc Natl Acad Sci USA* 105 (30): 10426–31.

Wilcox, C. 2016. *Venomous: How Earth's Deadliest Creatures Mastered Biochemistry*. New York, NY: Scientific American.

Willems, C., and S. Martens. 2016. Time to see the bigger picture: Individual differences in the attentional blink. *Psychon Bull Rev* 23 (5): 1289–1299.

Willems, C., J. D. Saija, E. G. Akyürek, and S. Martens. 2016. An Individual Differences Approach to Temporal Integration and Order Reversals in the Attentional Blink Task. *PLoS One* 11 (5): e0156538.

Williams, G. C., R. M. Nesse. 1991. The dawn of Darwinian medicine. *Q Rev Biol* 66 (1): 1–22.

Willis, R. J. 2002. Pioneers of Allelopathy. XII. Augustin Pyramus de Candolle (1778-1841). *Allelopathy Journal* 9 (2): 151–158.

Wilson, E. O. 1971. *The Insect Societies*. Cambridge, MA: Belknap Press.

Wilson, E. O. 2000. *Sociobiology: The New Synthesis*. Cambridge, MA: Belknap Press.

Wilson, E. O., J. M. Gómez Durán. 2010. *Kingdom of Ants: José Celestino Mutis and the Dawn of Natural History in the New World*. Baltimore, MD: Johns Hopkins University Press.

Wilson, E. O., B. Hölldobler. 2005. Eusociality: Origin and Consequences. *Proceedings of the National Academy of Sciences USA* 102 (38): 13367–71.

Winkelman, M. J. 2017. The Mechanisms of Psychedelic Visionary Experiences: Hypotheses from Evolutionary Psychology. *Front Neurosci* 11: 539.

Witt, P. 1954. Spider Webs and Drugs. *Scientific American* 191 (6): 80–86.

Witt, P. N. 1971. Drugs alter web-building of spiders: a review and evaluation. *Behav Sci* 16 (1): 98–113.

Wolf, M. E. 1999. Cocaine addiction: clues from *Drosophila* on drugs. *Curr Biol* 9 (20): R770–2.

Wolf, F. W., and U. Heberlein. 2003. Invertebrate models of drug abuse. *J Neurobiol* 54 (1): 161–78.

Wrangham, R. W., M. L. Wilson. 2005. Collective violence: comparisons between youths and chimpanzees. *Annals of the New York Academy of Sciences* 1036: 233–56.

Wu, J. P., and M. H. Li. 2018. The use of freshwater planarians in environmental toxicology studies: Advantages and potential. *Ecotoxicol Environ Saf* 161: 45–56.

Wu, Z., Y. Yang, L. Xie, G. Xia, J. Hu, S. Wang, R. Zhang. 2005. Toxicity and distribution of tetrodotoxin-producing bacteria in puffer fish *Fugu rubripes* collected from the Bohai Sea of China. *Toxicon* 46 (4): 471–6.

Wyatt, T. D. 2009. Fifty years of pheromones. *Nature* 457 (7227): 262-3.

Wyatt, T. D. 2010. Pheromones and signature mixtures: defining species-wide signals and variable cues for identity in both invertebrates and vertebrates. *J Comp Physiol A Neuroethol Sens Neural Behav Physiol* 196 (10): 685-700.

Wyatt, T. D. 2017. Pheromones. *Curr Biol* 27 (15): R739–R743.

Yamada, R., T. Tsunashima, M. Takei, T. Sato, Y. Wajima, M. Kawase, S. Oshikiri, Y. Kajitani, K. Kosoba, H. Ueda, et al. 2017. Seasonal Changes in the Tetrodotoxin Content of the Flatworm *Planocera multitentaculata*. *Mar Drugs* 15 (3): 56.

Yamamoto, D., K. Sato, and M. Koganezawa. 2014. Neuroethology of male courtship in *Drosophila*: from the gene to behavior. *J Comp Physiol A Neuroethol Sens Neural Behav Physiol* 200 (4): 251–64.

Yan, N., Y. Du, X. Liu, H. Zhang, Y. Liu, Z. Zhang. 2019. A Review on Bioactivities of Tobacco Cembranoid Diterpenes. *Biomolecules* 9 (1): 30.

Yokawa, K. et al. 2018. Anaesthetics stop diverse plant organ movements, affect endocytic vesicle recycling and ROS homeostasis, and block action potentials in Venus flytraps. *Ann Bot* 122 (5): 747–756.

Yokawa, K. et al. 2019. Anesthetics, Anesthesia, and Plants. *Trends Plant Sci* 24 (1): 12–14.

Yokawa, M. B., G. A. Mashour. 2019. The Biology of General Anesthesia from Paramecium to Primate. *Curr Biol* 29 (22): R1199–R1210.

Yoshida, A., A. Rzhetsky, L. C. Hsu, and C. Chang. 1998. Human aldehyde dehydrogenase gene family. *Eur J Biochem* 251 (3): 549–57.

Young, S. L. et al. 2011. Why on earth? Evaluating hypotheses about the physiological functions of human geophagy. *Q Rev Biol* 86: 97–120.

Yun, Y. et al. 2012. Sensitivity to silthiofam, tebuconazole and difenoconazole of *Gaeumannomyces graminis* var. tritici isolates from China. *Pest Manag Sci* 68 (8): 1156–63.

Zandawala, M., S. Tian, and M. R. Elphick. 2018. The evolution and nomenclature of GnRH-type and corazonin-type neuropeptide signaling systems. *Gen Comp Endocrinol* 264: 64–77.

Zars, T. 2012. Physiology. She said no, pass me a beer. *Science* 335 (6074): 1309–10.

Zer-Krispil, S., H. Zak, L. Shao, S. Ben-Shaanan, L. Tordjman, A. Bentzur, A. Shmueli, and G. Shohat-Ophir. 2018. Ejaculation Induced by the Activation of Crz Neurons Is Rewarding to *Drosophila* Males. *Curr Biol* 28 (9): 1445–1452.e3.

Zewde, A. M., F. Yu, S. Nayak, C. Tallarida, A. B. Reitz, L. G. Kirby, and S. M. Rawls. 2018. PLDT (planarian light/dark test): an invertebrate assay to quantify defensive responding and study anxiety-like effects. *J Neurosci Methods* 293: 284–288.

Zuckerman, B. et al. 1975. Detection of interstellar trans-ethyl alcohol. *Astrophysical Journal* 196 (2): 99–102.

# INDEX

# ABOUT THE AUTHOR

Dr. Oné R. Pagán is a husband and father as well as a biology professor, scientist, blogger, podcaster, and author. He has published original work in various scientific journals including the *International Journal of Developmental Biology*, *Neuroscience Letters*, the *European Journal of Pharmacology*, *Toxicon*, *Neuroscience*, *Philosophical Transactions of the Royal Society B*, *Neurochemical Research*, and *Pharmacology, Biochemistry, and Behavior*, among others. He holds an undergraduate degree in natural sciences and a master's degree in biochemistry, both from the University of Puerto Rico, and a doctorate in pharmacology with an emphasis in neurobiology from Cornell University.